MW01181928

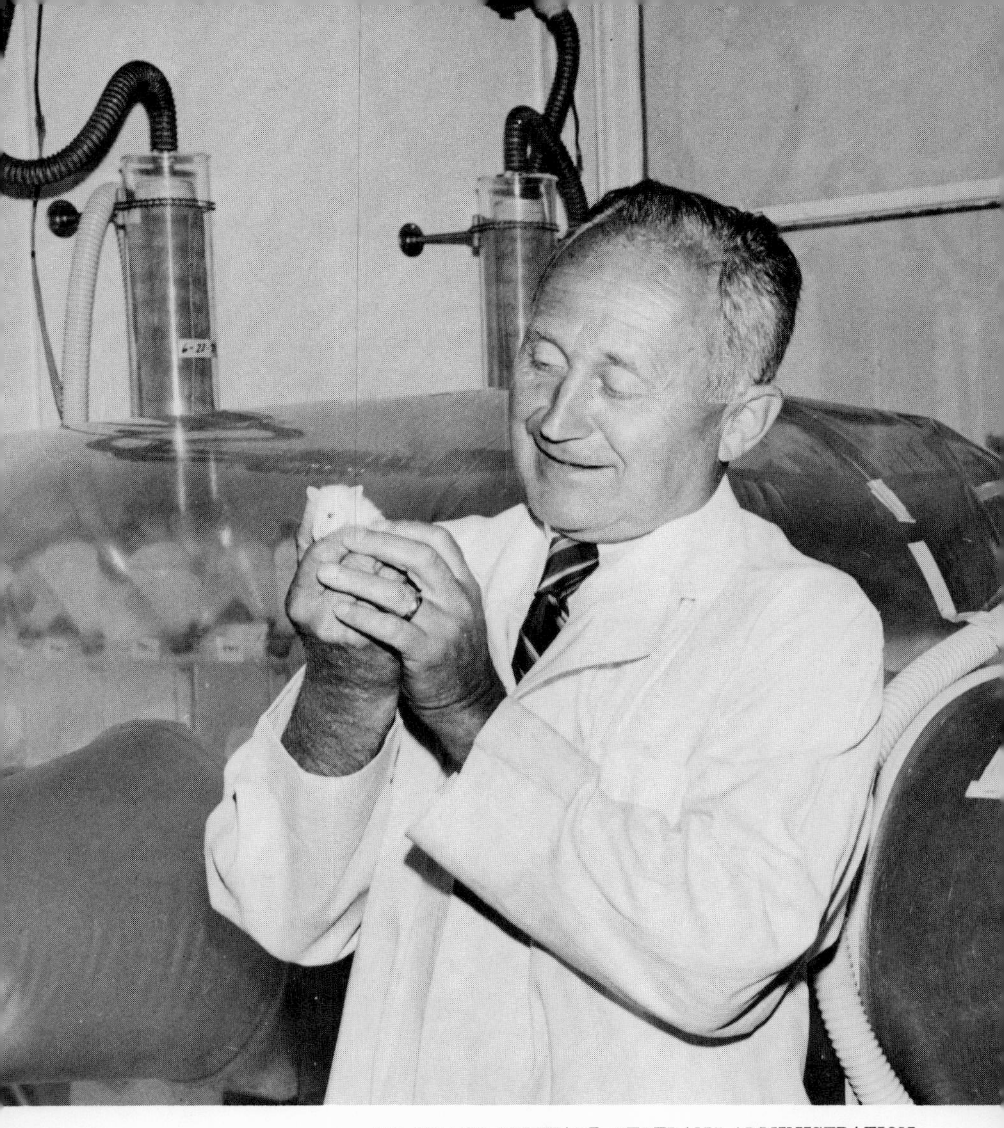

DR. KLAUS SCHWARZ, VETERANS ADMINISTRATION

Dr. Klaus Schwarz is one of many scientists finding out how various trace elements help us and harm us. Here he is shown with a mouse for research and a sterile isolator that he developed, in which test animals are kept free of all contaminants.

TRACE ELEMENTS

How They Help and Harm Us

Joan Arehart-Treichel

HOLIDAY HOUSE / NEW YORK

To Horst and Marguerite

Printed in the United States of America

LIBRARY OF CONGRESS CATALOGING IN PUBLICATION DATA

Arehart-Treichel, Joan.
Trace elements.

SUMMARY: Discusses the origins of various elements and compounds and their beneficial and detrimental effects on animals, plants, and humans.

1. Trace elements—Physiological effect—Juvenile literature. 2. Plants, Effect of trace elements on—Juvenile literature. [1. Trace elements—Physiological effect] I. Title.
QH545.T7A73 574.1'921 73-16875
ISBN 0-8234-0242-8

Contents

1/ The Origin of the Elements

Have you stood on a hill on a crisp autumn night to study the sky overhead? Did you wonder where the stars came from? How the earth and her neighboring planets were born? If so, you wondered what people have wondered for thousands of years—how the universe was set into motion.

Scientists who study the stars and planets—cosmologists and astronomers—know what stars and planets are made of. They are composed of material called elements. Elements are made of atoms. Individual atoms are so small that we cannot see them with our naked eyes. Scientists can see some of them, but only with powerful instruments. Each atom contains still smaller units of matter known as electrons, protons, and neutrons. Each element is distinguished from the others by its particular weight. An element's particular weight depends on how many electrons, protons, and neutrons the element contains.

Some 98 elements are known to be naturally present in stars and planets, and also in the clouds of dust and gases that swirl among them. Cosmologists (scientists who study the entire universe) and astronomers are convinced that all these elements first came into being billions of years ago. And as elements were made, stars and planets were born. Before there were elemental particles and elements, the universe was nothing —a silent, icy-cold realm.

Scientists have studied stars and planets for several thousand years at least. But they are still not sure how elements were made and how stars and planets evolved from them. Still, modern scientific equipment and ways of asking questions of nature have enabled them to get some good ideas about the puzzle. Two explanations are popular at present. The more popular one is the big-bang theory.

This idea says that the universe started with a dense mass of neutrons, particles with no electrical charge. As this cloud of elemental particles whirled in cold, black space, it came together. This coming together, or condensation, of matter, created heat. The heat became so great that the tightly packed ball of neutrons became a fireball. The ball spun around and around and then burst with a big bang. In perhaps as little time as half

NATIONAL RADIO ASTRONOMY OBSERVATORY

The galaxy Messier 51. From such "celestial pinwheels" of matter stars and planets eventually form.

an hour, all the elements were made. The elements then cooled and condensed into galaxies, or celestial pinwheels of elements. Out of the galaxies stars and planets formed.

The other theory is the steady-state theory. This idea says that the elements were not made in 30 minutes or even 30 days, but over millions, even billions of years. The elements evolved as stars evolved.

This theory suggests that the universe started with a cold, swirling mass of hydrogen atoms. Some of these atoms condensed and formed stars. The hydrogen stars roamed space for a long, long time. But little by little their centers grew dense and hot. The heat was so fierce it changed the atomic makeup of the hydrogen.

So this gas formed a new, somewhat heavier element, helium. Helium is also a gas, the second-lightest element known.

Then as time went by, the stars that contained hydrogen and helium gradually condensed and became hot. These stars produced a still heavier element—perhaps carbon. Then the carbon interacted with the helium to form even heavier elements—oxygen, say, or magnesium. When the helium was used up, the stars contracted. They again became hot—hot enough to make carbon combine with oxygen and magnesium. As a result, still heavier elements such as lead and bismuth formed. It probably took as long as one or two billion years for the heavy elements to be made.

Shrinking, Exploding Stars

As time went by, many stars grew old and shrunken. They exploded. And the elements making up these stars shot out into space like missiles. These elements were then able to form new stars, and eventually planets. Slowly but surely the simple hydrogen became the action-packed universe.

The birth and death of stars is as old as the universe. But it is also taking place today. Some of those stars winking at you from the night sky may be dying or being born. The result of

MOUNT WILSON AND PALOMAR OBSERVATORIES

The Crab Nebula consists of the matter shot out by an exploding star, in this case a supernova that exploded in 1054 A.D.

this death and birth is a continual reshuffling of elements.

By looking through telescopes and listening to radio waves that stars send in all directions, cosmologists and astronomers know that some stars are giving off phosphorus, chlorine, manganese, carbon, and other elements. These will probably become part of new stars. How dying stars eject elements is not understood well, though. And how these elements become new stars is also largely a mystery, in spite of hard work by the investigators.

The more scientists learn about the origin of the elements, and the composition of stars and planets, the more certain they are that the first cosmological events took place a very long time ago. They used to think that the elements were formed five billion years ago. This is a span of time that is almost beyond our wildest imagina-

The 150-foot radio telescope at Stanford University in Stanford, California. Such instruments help tell scientists what elements are being emitted by various stars.

STANFORD UNIVERSITY

tion. Now they are more inclined to think that the elements may have been formed even 10 or 20 billion years ago.

On the basis of information obtained from telescopes, spaceships, and space probes, they are fairly confident that the star closest to us, the sun, was made about five billion years ago. The earth and our neighboring planets Mercury, Venus, Mars, Jupiter, Saturn, Uranus, Neptune, and Pluto, were probably formed shortly after the sun was born—say, 4.6 or 4.7 billion years ago. The oldest rocks we have found on earth are in Greenland. They date back four billion years. The precious rocks brought back from the moon by American astronauts date back 3.6 to 4.5 billion years, so the moon must have been born about the same time as the earth and other planets.

Most peculiarly, samples of soil from the moon date back 4.2 to 4.9 billions of years. How can the soil possibly be older than the moon and the rest of our planetary system? Scientists suspect that the soil is probably not older, only that their techniques for dating soil are not altogether accurate. So there is a lot of controversy about the age of lunar soils.

As more is learned about the amounts of elements in stars, planets, and the dust and gases in space, the more amazed scientists are by the proportions of them. About 98 per cent of stars and interstellar dust and gas is made of the two lightest elements, hydrogen and helium. That leaves only 2 per cent to be made of the other known elements. Yet these many elements are widespread throughout the universe. Dust clouds, for example, have been found to contain abundant amounts of phosphorus, chlorine, manganese, carbon, nitrogen, oxygen, magnesium, silicon, sulfur, iron. Meteorites—chunks of planets shooting through space—are known to contain the elements selenium, arsenic, antimony, tin, mercury, zinc, gold.

Planet and Moon Elements

Cosmologists, astronomers, and geochemists (scientists who study the substances in the earth) are fairly well acquainted with the elements in the earth, our moon, and our neighboring planets. But they are not at all sure why they contain the elements they do.

The cores of the larger planets, Jupiter, Saturn, and Uranus, are made almost entirely of the light elements hydrogen and helium. You would think that because they are so gigantic, they would be made of heavier elements. On the other hand, the cores of the smaller planets, Venus, Mars, Mercury, our earth and moon, are made mostly of heavier elements, especially iron and silicon. The earth's innermost core, like most of

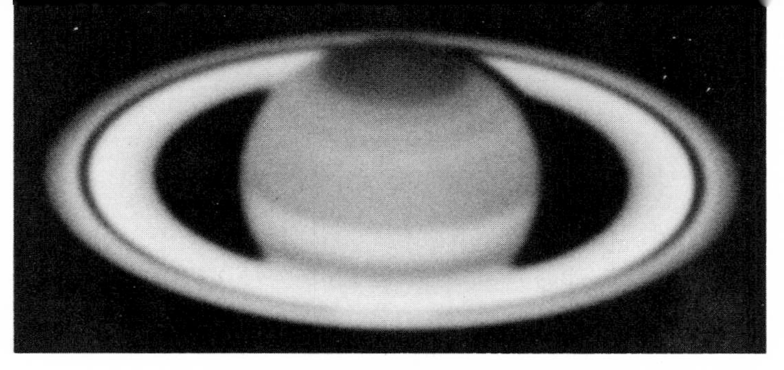

MOUNT WILSON AND PALOMAR OBSERVATORIES

Saturn is one of the planets now known to have cores of very light elements rather than heavy ones.

its outer core, is iron. But if you could pick up this innermost core, it would have the faintest bit of "give," like a ball of lead.

The investigators think that the earth, our moon, Venus, Mars, and Mercury contained light elements as well as middle-weight and heavier elements when they were first formed. But the light elements were quickly blown into space by a constant stream of charged elementary particles from the sun—the so-called "solar wind." Such an event would have been possible. These planetary bodies are much closer to the sun than are Jupiter, Saturn, and Uranus. Also, the light elements were held by the great gravitational pull of the larger planets.

The Apollo 16 spaceship trip to the moon brought back samples of moon rocks. When geochemists examined the elements in these rock samples, they found various kinds. But there were hardly any of the light elements carbon, nitrogen, hydrogen, and oxygen in the rocks. They concluded that these were left behind in the sun when the moon formed.

Today, only light elements make up the atmospheres of the earth and our planets. This makes sense, since any middle-weight or heavy elements have been pulled down to the surfaces of these bodies.

The earth's atmosphere is rich in hydrogen, carbon, and oxygen—or more specifically, in carbon dioxide (carbon plus oxygen); oxygen molecules (two atoms of oxygen per molecule); and water (hydrogen plus oxygen). Oxygen, hydrogen, and carbon have been detected in the atmosphere of Mars and Venus. These elements are present mostly as carbon dioxide with only traces of water. The moon has hardly any atmosphere, which means it has hardly any elements floating around it. The only exception is traces of the element argon, which probably came from the sun. The reason the moon has almost no atmosphere is that it is a tiny celestial body, much smaller than the earth or our neighboring planets. As a result it does not have enough gravitational pull to hold elements in space near it.

A spaceship now on its way to Jupiter and back—a six-to-ten-year journey that began in 1972—will tell us how much hydrogen and helium are in the atmosphere of that distant, intriguing planet. The spaceships will also say how much of Jupiter's atmosphere contains methane (carbon plus hydrogen), ammonia (nitrogen plus hydrogen), and water vapor (hydrogen and oxygen).

One of America's leading cosmologists says: "Jupiter today may have the kind of atmosphere that the primitive earth had four to four-and-a-half billion years ago."

2/ *Elements Become Life*

Two thousand years ago, the Greek philosopher and scientist Aristotle suggested that life might have formed on the primitive earth from earth, air, fire, and water. Even today, scientists who study the elements that make up life—biochemists—think Aristotle was probably right, loosely speaking.

Six elements make up the bulk of living things. They are all light elements. They are hydrogen, carbon, nitrogen, oxygen, sulfur, and phosphorus. These were probably present in the atmosphere of the ancient earth, in its crust and in its oceans. The elements could have been made to interact, to form chemical compounds, by powerful rays—radiation—pouring from the sun. Or the elements might have come together under the force of lightning.

Biochemists have attempted to recreate primitive earth conditions in the laboratory.

UNIVERSITY OF CALIFORNIA

Dr. Stanley Miller made a substantial step forward by recreating in the laboratory the chemical conditions that existed on the earth very early in its history. Here he is shown with his apparatus, in which he produced sparks (to imitate lightning) above certain chemicals that were present in the waters of the earth at that time.

They have put methane, ammonia, and water together and have bombarded them with electrical discharges. The electricity imitated the radiation or lightning that may have triggered the first formation of chemical compounds on earth. In fact, some of the chemical compounds that today make up life have been made under such laboratory conditions.

Hydrogen and carbons have joined as hydrocarbons. Hydrocarbons have come together with oxygen to make acids. Acids have joined with more oxygen to make sugars. Hydrogen, carbon, and oxygen have come together to make proteins.

Not all life compounds, however, have been recreated under laboratory conditions. All of them probably never will be. The coming together of elements to make the compounds of life must have been incredibly complicated. Compounds were probably formed over millions of years and under ever-changing earth conditions.

As far as biochemists can tell, there are several reasons why light elements on and near the primitive earth became the building blocks of life. One is that these light elements were plentiful, just as they are plentiful throughout the universe. Another reason is that light elements combine easily with each other. Light elements also make strong, stable chemical compounds. Heavier elements do not do so well in linking up with other elements. And when heavier elements do combine with other elements, the resulting compounds are not always sturdy.

"Soups" For Life

Some biochemists think the compounds that were eventually to become life were made in the oceans of the young earth. These oceans probably bubbled and hissed like steaming bowls of soup, and biochemists in fact refer to them as "primordial soups." Because the soups were active cauldrons of elements, life-forming compounds may have been made in them.

Other biochemists, however, do not think this happened. They believe that life-forming compounds were made in the primitive air and on the primitive earth crust rather than in the oceans. The reason they disagree is that they performed laboratory experiments in which they put the light elements that make up life into imitations of primordial soups. The light elements formed few life-forming compounds in the water. So it looks as if life-forming compounds were not made in the oceans, but elsewhere.

Most biochemists do agree, though, that once the chemical compounds that make life were formed, life did begin in the primitive oceans. Quite a number of laboratory experiments support this view. For example, when several primitive forms of life chemicals were made in the laboratory under ancient-earth conditions, water was needed. Also, living things today are made largely of water. This suggests that life arose from the seas.

No one knows when the first spark of life appeared. But the first forms of life were probably similar to the basic units of life today. These are cells—tiny packets of chemical compounds. In most cases we cannot see cells with our naked eyes. But they are easy to see under a microscope.

Two things make cells different from the nonliving chemical compounds of which they are made. One is that cells carry out various activities—eating, breathing, repairing themselves if they are hurt, reproducing. Another is that cells need continual supplies of elements and compounds to carry out these activities.

The First Living Cells

The oldest record that geochemists have of the first cells—the first forms of life—on earth consists of imprints in rocks of one-celled plants, something like modern blue-green algae. These plants appear to have lived about 3.3 billion years ago, although plants may have lived before that. This means that the first forms of life appeared on earth only a half-billion years after the earth itself was formed. In short, life is very, very old.

The fossil imprints in the rock also suggest

that those early plants used light elements for cellular activities—photosynthesis, for example. Photosynthesis is a process of most plants by which they make their own food. They take carbon and oxygen from the air (in the form of the compound called carbon dioxide) and hydrogen and oxygen from water. With the help of sunlight they turn these elements into sugar. The sugar gives plants energy. During photosynthesis—that is, while reacting to light—plants also give off oxygen into the air as a waste product. (Their actual respiration, which is a different process, is chemically much like our own.)

No one knows when one-celled animals appeared on the ancient earth. But scientific discoveries during the 1960s seem to show that they probably existed as early as 3.3 billion years ago.

When did one-celled plants and animals evolve into many-celled plants and animals? There is evidence that many-celled organisms lived at least 1.9 billion years ago. They too were made of chemical compounds. These compounds in turn were made of the six light elements carbon, hydrogen, nitrogen, oxygen, sulfur, and phosphorus. They took these elements from the air, earth, and oceans to feed themselves, to breathe, to repair themselves, and to reproduce. Plants that reproduce by seeds probably did not appear until the amount of oxygen in the atmosphere was close to what it is today.

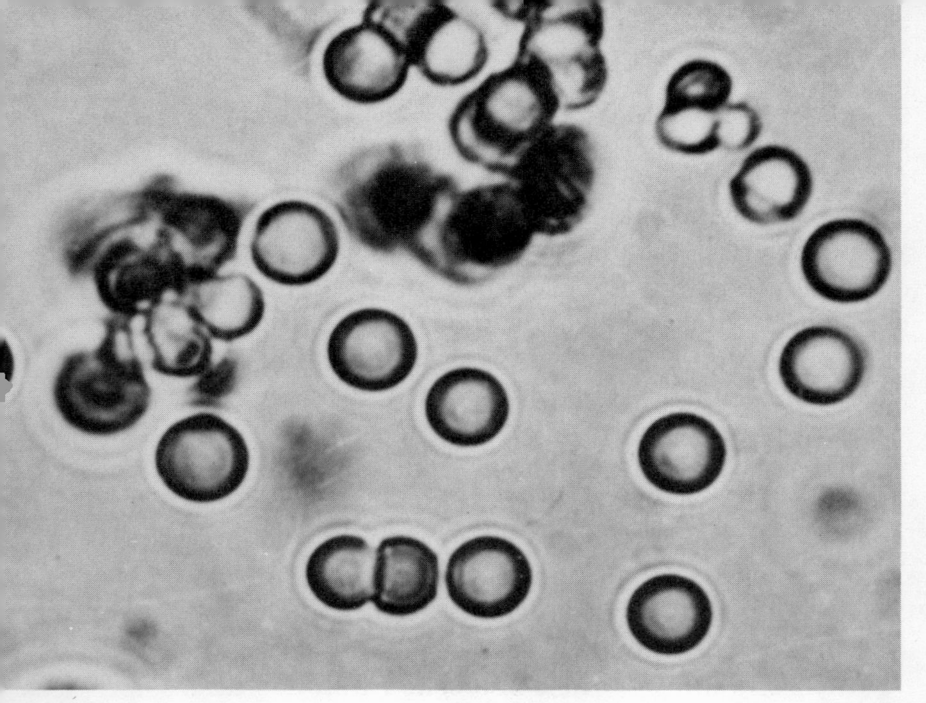

DR. SIDNEY FOX, UNIVERSITY OF MIAMI

These "microspheres" were made by the reaction with water of protein-like substances such as were found on the early earth, at high laboratory temperatures. Each one is about two-thousandths of a millimeter wide. They are not living cells, but they suggest the way the first living cells may have formed.

Humans Appear

Then, somewhere around two million years ago—which is only a sliver of time compared to the time it took for stars, planets, plants, and animals to evolve—people appeared on earth. They too were made of many cells. Like plants and animals, they needed supplies of light elements to feed themselves, to breathe, to stay healthy, and to reproduce. Plants got the elements they needed from the air, earth, and water. Animals got the elements they needed from the air, earth, water, and plants. People got the elements they needed from the air, earth, water, plants, and animals.

So the universe evolved. Plants, animals, and people were made on earth. And all this was possible because of the elements, especially because of a few light elements. Hydrogen is a major part of stars, and a major component of life as well. So the saying "We are the stuff that stardust is made of" may be more than fantasy.

Our lives, and those of other living things, though, do not depend just on the six light elements hydrogen, carbon, nitrogen, oxygen, sulfur, and phosphorus. Smaller amounts, or traces, of other elements can also be found in plants, animals, and people. How did traces of other elements come to be a part of life? Biochemists have looked at modern cells for clues.

They know that other elements than the major six themselves are present in our cells in small amounts. They know that these "trace elements" help cells breathe, eat, stay healthy, and reproduce. They believe that the same trace elements were also needed by the first cells on earth, and for the same reasons, though there is little scientific evidence for this so far.

One recent laboratory experiment, though, does offer some support for this belief. A kind of reproductive primitive "cells," though not actually alive, was made under laboratory conditions like those of the primitive earth. Acids came together to make this "cell." But they did so with the help of heat and of small amounts of sodium and magnesium. These are two of the elements that are needed by present-day cells in small amounts. So it looks as if traces of these same elements were useful to life from the beginning.

3/ *Trace Elements Help Life*

Plants, animals, and people have probably contained bits of various elements for millions of years. But biochemists have discovered trace elements in living things only during the past 50 years or so. The reason is that they needed modern scientific instruments to detect such tiny amounts of chemicals in living tissues.

Trace elements do some very important things. For one, they glue a living thing together, so to speak. When many cells are grouped together in a living organism, the unit of cells is known as a tissue. When tissues come together they may make an organ. Organs and tissues that work together in an organism are known as a system. The more complicated an organism is, the more tissues, organs, and systems it will have. But whether an organism is a one-celled bacterium or a person with some 180 billion cells,

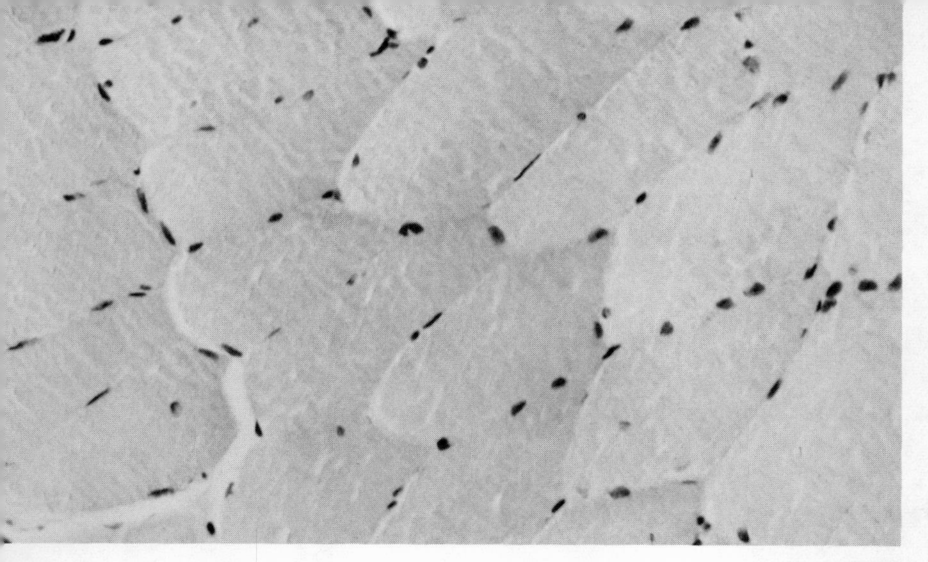

LEWIS E. KOSTER

Muscle cells are shown here grouped into bundles of muscle fibers (each egg-shaped mass is a fiber group); such fibers make up a muscle, one of seven main types of tissue in animal and human bodies.

trace elements play important roles in holding its parts together and ensuring that they work together.

The other important role that trace elements play in plants, animals, and people is as helpers called enzymes. These enzymes are proteins, but a special kind of protein. They help cells change various chemicals into compounds that cells, and the entire organism, need in order to eat, breathe, make repairs, and reproduce. An enzyme helps cells carry out these reactions without being changed itself. We might say it is like the trigger on a gun—it gets the bullet going but it stays where it is on the gun mechanism. Trace elements may help out as a basic part of an enzyme or, in a certain form, give a boost to the enzyme from outside.

Trace elements that help life appear to be mostly middle-weight elements. One reason is that these elements are fairly common in the air, the crust of the earth, and the oceans, but not so common as the six light elements that mostly make up living things. Heavier elements are not so common. Another reason is that middle-weight elements are fairly "friendly" toward other elements. Under the right conditions they are able to interact with light elements and give matter stability. They help enzymes speed up chemical reactions but do not form permanent bonds with chemical compounds. Heavier elements, on the other hand, seldom form bonds with other elements. Therefore they are not ideal candidates as enzyme assistants. But traces of a few heavier elements may be important to life. We will look at them later on.

Whether an organism uses smaller amounts of this or that element depends on two things. For one, how plentiful the element is in the surroundings of an organism. For another, on the organism's particular needs.

The Oyster's "Preference"

Take the oyster. The light elements that make up much of life—carbon and oxygen—combine with middle-weight elements, such as calcium or strontium, to make the bones and shells in various animals. But for some reason, the oyster uses carbon, oxygen, and magnesium to make its shell. Why is this? As far as biochemists can figure out, magnesium is used by the oyster because magnesium is plentiful in the ocean. But there is a lot of strontium in the ocean too; so why does the oyster use magnesium instead of strontium? Apparently because magnesium is more stable, a somewhat lighter element than strontium. But the oyster's body uses traces of strontium also, to make hinges between the valves of its shell. Obviously these hinges have to be flexible. Strontium, being a rather heavy and somewhat unstable element, provides this flexibility.

Whether an animal or plant uses small amounts of this or that element, then, depends on whether the element is there in its environment, and on its particular biological needs.

What have we learned so far? Light elements make up the bulk of living matter. But middle-weight elements, and some heavier elements, form small supporting links in living matter. Traces of middle-weight elements, and a few heavier elements, help enzymes turn light elements into living matter. These enzyme helpers assist each cell, and in the end the whole organism, to make the chemical compounds they need for eating, breathing, repairing themselves, and reproducing.

Trace elements are a small aspect of life. But they are extremely important and exciting. In fact, as we shall soon see, no plant or animal, neither you nor I, could survive without them.

4/ Helping Plants To Live

For centuries people had been mystified by how plants live, grow, and reproduce. What chemicals are plants made of, they asked. Where do plants get these elements? Actually, it has been only a hundred years or so since scientists who study plants—plant biologists, botanists—have solved the mystery.

Plants, like other forms of living matter, are made almost entirely of several light elements. They get carbon and oxygen (as carbon dioxide) from the air. They get hydrogen and oxygen (as water) from the soil; likewise phosphorus and sulfur. They take nitrogen from the soil, and possibly from the air as well. Like animals, plants build up these light elements into proteins and other compounds that they need to live a plant's life.

During the past 50 years, plant biologists have also discovered that plants are made of traces of other elements—very important ones. In 1920, 10 elements were known to be needed by plants. These were the light elements carbon, hydrogen, oxygen, nitrogen, phosphorus, and sulfur, and traces of the middle-weight elements potassium, calcium, and magnesium. Now other trace elements are known to be important too—these include iron, manganese, boron, copper, vanadium, molybdenum, chlorine. Trace elements are needed by *all* plants, from the one-celled alga that lives in water to the towering redwood trees of the West.

One of the most important trace elements for plants is magnesium. It is part of the pigment that makes leaves green. This pigment is chlorophyll. It is used by leaves to make food and to breathe—a process called photosynthesis. The billions of little particles of chlorophyll in the leaves and stems of a plant convert carbon dioxide in the air, and water taken up from the soil, into sugar, and return oxygen to the air. Without magnesium there would be no chlorophyll. Without chlorophyll there would be no photosynthesis. Without photosynthesis a plant could not eat. Magnesium is vital for plant life—and for our life, since we depend on plants.

Iron is not part of chlorophyll, but it *is* part

BROOKHAVEN NATIONAL LABORATORY

The leaves of almost any plant are filled with chlorophyll, which manufactures the sugars it needs for nourishment. Without magnesium this sugar-making could not take place, for this trace element is a necessary part of the chlorophyll.

of the enzyme that helps a plant make chlorophyll. Without iron, magnesium could not be put into chlorophyll. Without chlorophyll a plant would starve. Therefore iron, like magnesium, is absolutely necessary for plants. Iron, in fact, is the most abundant trace element found in plants. Spinach, carrots, radishes, and clover especially have a lot of it.

Manganese is a helpful enzyme in many plants. This trace element helps plants turn nitrogen into various chemical compounds that they need. Like iron, manganese helps enzymes make chlorophyll in plant leaves.

Most trace elements that are important for plants are middle-weight elements. But there is one light element that is also important: boron. It helps make some plant proteins. Boron also seems to help plants get water from the soil, and to help enzymes convert light elements into plant food.

Plants also need traces of copper, molybdenum, and chlorine. Chlorine is needed by plants in larger amounts than any other trace element except iron. Traces of selenium may help plants. So may vanadium—two species, a mold and a green alga, are known to need vanadium. Aluminum helps sunflowers and corn grow. Aluminum makes soil acid. Some plants, such as mountain laurel, thrive on acidic soil. If you add aluminum to the soil around a hydrangea bush, its fluffy pink flowers will turn a deep blue.

Into the Soil and Out of It

Plants get the trace elements they need from the soil. Rocks in the earth's crust contain 80 of the known elements. Wind, rain, snow, drought, and other weather factors erode rocks, wear them away. With erosion, tiny particles of elements pass from rocks into soils.

When rain leaks down into soils, trace elements mix with this water and may form chemical compounds with it. Such chemical

USDA

The erosion of rocks and soil digs out small particles of trace elements and distributes them into the soil over large areas by means of running rainwater.

changeovers may affect how much of the trace elements plants are able to absorb. Microorganisms, especially bacteria and fungi, may play important roles in making trace elements in the soil available to larger plants. Bacteria and fungi can also help erode rocks so that more trace elements mix with soils. Fertilizers used on soils can alter the amounts of trace elements that are available to plants. In one scientific experiment, cow manure increased the amount of copper in the soil. But a different fertilizer, peat, decreased the amount of copper.

Soils vary greatly in the matter of *which* trace elements they contain. They also vary in *how much* of each there may be. Some trace elements form compounds with other elements in the soil very easily; others do not. These "preferences" can influence which trace elements plants absorb, and how much. For example, if vanadium hooks up with lead, zinc, and copper in soil, it may be hard for plants to absorb vanadium.

However, the amounts of trace elements that plants take in through their roots vary much less than the amounts of trace elements found in soils. As the elements move from the surface of the root up the plant stem for use in plant leaves, all kinds of complicated reactions go on inside the plant. Plant species differ in their uptake of trace elements—even plants of the same species can vary in their use of the same elements.

If soils contain enough trace elements and they are in the right form, plants will soak up all they need. But if soils are poverty-stricken in usable trace elements, the plants will show signs of trace-element deficiencies, just as they show signs of other diseases.

Take zinc. The importance of zinc to plants was first noted in the nineteenth century. A French plant biologist discovered that zinc is necessary for the growth of a particular fungus. A few years later it was found that zinc is necessary for other plants as well. In 1927 farmers in Florida reported that vegetable crops grew better if zinc was added to the soil. In the 1930s, California fruit growers found that zinc im-

BROOKHAVEN NATIONAL LABORATORY

Every species of plant uses trace elements in the particular amount necessary for it. Even different plants of the same species may vary in their use of trace elements.

proved the growth of citrus fruit crops—oranges, tangerines, grapefruit. Zinc deficiency among plants has increased spectacularly since about 1953. Nobody knows quite why. Soil conditions and farming methods may have changed the amount of zinc in soils.

When a plant suffers from zinc deficiency, it grows well, but is not able to reproduce right. Corn grown in zinc-deficient soil will grow a lot of the tassel, or silk, but it will not develop ears of corn on its stalks. Without zinc, peas may bloom but they will not seed. A patch of zinc-deficient beans may yield only partial crops.

While there is a great deal of iron in soils, it seems probably the trace element in which plants are most deficient. When plants do not get enough iron, their leaves become spotty. If the deficiency is severe the chlorophyll in the leaves may fall apart. As a result the leaves lose their green color. Iron deficiency is one of the easiest trace-element deficiencies to see in plants. Pines, pin oaks, blueberry bushes, and azaleas are some of the plants that suffer the most if they do not get enough iron. Crops on farms in the West of the United States often suffer from iron deficiency.

The amount of manganese in plants probably varies more than that of other important trace elements. If cauliflowers are hungry for manganese they crinkle and wilt. Tobacco leaves starved for manganese turn pale. If oats and soybeans do not get enough manganese, they turn gray at the base of their leaves. Many states have reported soils that do not have enough manganese for plant needs.

Because plants can use only part of the light element boron as a trace element, they must have a continual supply of it from the soil. But many soils do not have enough of this element for the needs of plants.

Copper is rare in many soils. Fortunately plants do not need very much of it. If onions fail to get enough copper to meet their trace-element needs, they become yellow and mushy.

Plants need large amounts of chlorine. But soils do not always have enough chlorine for their needs. Fortunately many soils are enriched with the chlorine in sea water. Sea spray is blown onto the soil from oceans, traveling about 25 miles on the average, though under special weather conditions it may reach as far inland as several hundred miles.

5/ *Trace Elements at Work in Animals and People*

Elements in very small amounts have probably been valuable to the health of animals and people for thousands of years. But only since the 1950s has great progress been made in understanding what they do. Some trace elements play only one or two roles, but these are important indeed.

Iron was the first element found to be helpful. This was first noted in the seventeenth century. About 70 per cent of the iron in a healthy person is in his or her blood. It makes up part of the protein pigments that in turn make hemoglobin. This chemical in turn makes up part of red blood cells, which are a large part of our blood. Hemoglobin, particularly the iron-containing pigments in it, are what make blood red.

Iron and Copper In Us

Iron, so plentiful in the earth's crust, is a precious element in the human body. There is only enough in each of us to make one small iron nail. After a red blood cell has spent some 120 days moving throughout the body, the iron in the cell is released from its hemoglobin, is saved by the body and is used once again. Little iron is excreted in urine; we humans are iron misers.

If a person is anemic, he does not have enough red blood cells. Or those cells do not have enough hemoglobin. Anemia can be caused by different things—for example, if you lose enough blood in an accident or if you do not get enough iron from the foods you eat. People with anemia are often tired, pale, and irritable.

A lot of people, particularly children and women, are anemic because they do not get enough iron in their diet. Iron is not particularly common in foods. What's more, the iron in many foods is not easily digested and made available to the body.

Red meats are good sources of iron. Other foods that are fairly good sources include chocolate, peas, molasses, parsley, and dandelion leaves. If your mother says that spinach will give you lots of iron, you can explain to her that spinach does contain a lot of iron, but much of this is not absorbed by the human body. In other words, spinach may be a valuable source of some vitamins and minerals, but not of iron, particularly.

The best source of iron for children and

other people, nutrition scientists say, is a daily vitamin-mineral pill that contains quite a bit of iron. Also, if your mother cooks foods in iron skillets or pots, traces of iron from these utensils will combine with foods cooked in them. This iron can supplement your daily iron intake.

Copper is also important to animals and people in trace amounts. Copper was used to treat various human ailments by the Egyptians 1500 years ago. The Greeks prescribed it for lung diseases and mental disorders. By the eighteenth century copper was used widely as a treatment. But its medical popularity started to decline around 1875. This decline was rather strange. The popularity of this metal for treatment should have increased about this time, if anything, because the role of copper was beginning to be understood. Copper was known to combine with blood proteins in snails. It was found to be a normal part of human blood, and was related to the transport of oxygen.

In 1924 biochemists in Wisconsin showed that rats and rabbits raised on a diet of milk developed anemia. The animals did not get better when they were given iron treatments. But they did improve after they were given copper. More experiments followed and confirmed this exciting discovery. By 1935 scientists were positive that copper is important to blood.

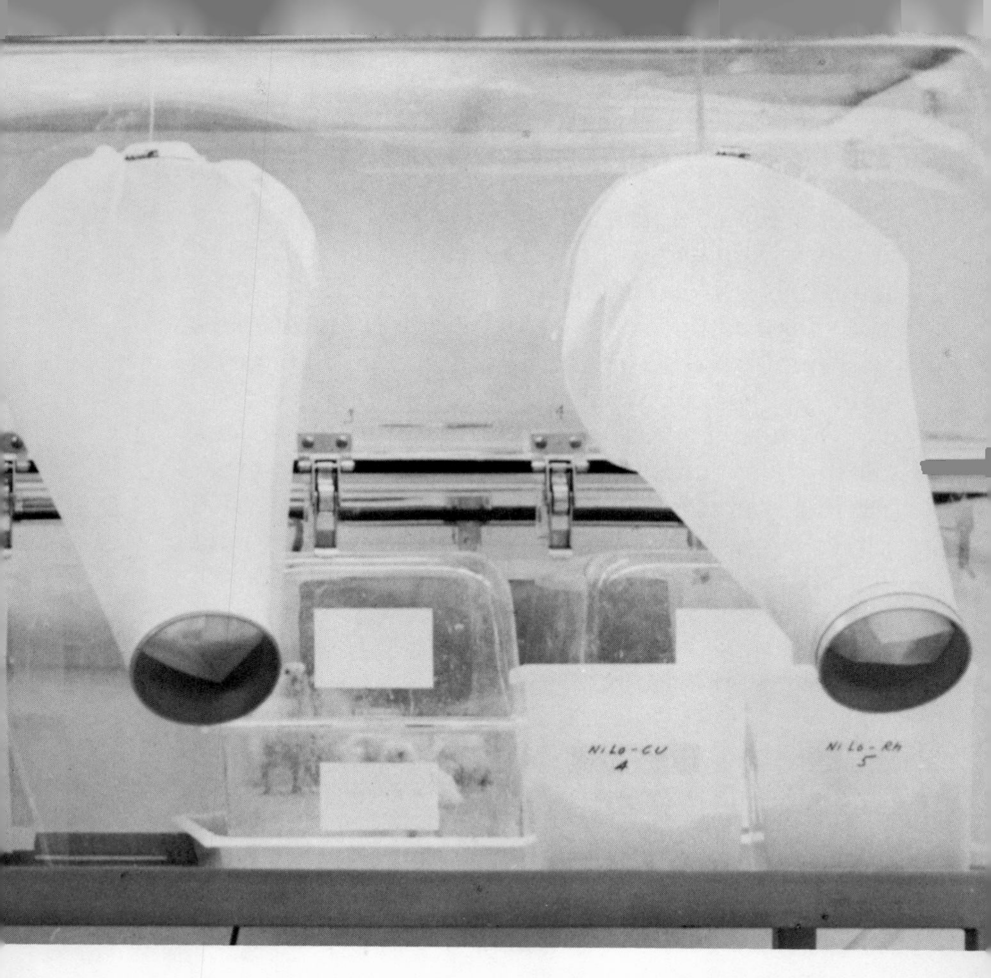

HUMAN NUTRITION LABORATORY,
AGRICULTURAL RESEARCH SERVICE,USDA

Researchers are busy all over the country confining test animals, like these chicks, in isolators which keep them free of accidental trace elements in their environment. In this Department of Agriculture experiment by Dr. Forrest H. Nielsen, contaminating dust is removed from the air supply by the baglike filters. The chicks are then exposed to the specific trace elements being studied.

Today we know that copper is present in all animals and people. Before a baby is born, the copper in its liver increases until there is five to ten times more copper than in the adult liver. Once a baby is born, this copper decreases rapidly during the first year of life. The only organ in the human body that is known to accumulate more copper with age is the brain. Your grandmother or grandfather probably has twice as much copper in her or his brain as you have in yours.

There are certain tissues and organs in all animals and people that contain more copper than other tissues and organs do. These include the liver, brain, heart, kidneys, and hair. The eyes of many animals contain copper. Women have more copper in their blood than men do.

More than 20 copper-containing proteins are now known. They are widely distributed in animal and human tissues. Biochemists are not sure how many of these proteins are pigments and how many are enzymes. Because copper is an element that combines fairly easily with other elements, it is possible that most of these copper proteins are enzymes. The copper in other words, would be suitable for helping enzymes bring about various changes and processes. These reactions could include growth and maturation of cells. Copper seems to be necessary for the growth of red blood cells, for their becoming mature, and for their movement about the body.

Protein Chariots

Copper is carried throughout the body by proteins, to which it is loosely hooked up, in the bloodstream. It then leaves the blood to move into tissues and organs that need it. It seems that there are certain sites on the cells in these tissues and organs. These sites attract copper atoms and either store them or release them into certain enzymes in the cells.

How copper leaves organs and tissues to return to the bloodstream and eventually pass into the urine is not understood very well. Animals and people vary considerably in the amount of copper they excrete daily in their urine. There is also no connection between the amount of copper you take in through your food each day and the amount of copper that passes out of it.

Many illnesses have been linked with a diet that is deficient in copper. They include anemia, bone disorders, a lack of growth, heart failure, stomach upsets, a loss of pigments in hair or wool. Whether a person or animal suffers from a copper deficiency depends on how bad the deficiency is, how old the person or animal is, and other factors. Copper deficiency in people has been a subject of controversy almost as long as copper has been suspected of being an important nutrient.

A few years ago, for example, a group of physicians tried to find out whether people with iron-deficiency anemia respond better to copper therapy added to iron therapy than they do to iron therapy alone. The results were confusing. But since then, other scientists have found that people can definitely suffer from lack of enough copper. Some people who have a deficiency do not get enough copper in their diets; others have a disease they were born with that robs their bodies of their copper. Babies kept on an all-milk diet for a long time after birth can also suffer from a lack of copper.

As we said earlier, plants, especially leafy vegetables, can suffer from a soil that hasn't enough copper. But copper deficiency in people has never been traced to any particular geographic area where plants are copper-poor. The reason is probably that most people today get their foods from many different parts of the United States. In fact, scientific evidence suggests that few Americans suffer from copper deficiency. Adults need only a little of this metal in their diets every day. Children need even less. Sources of dietary copper include fish, meat, eggs, nuts, beans, cocoa. One baby doctor wanted to make sure that baby foods contain enough copper for the needs of infants. She sampled

THE ANACONDA COMPANY

Copper is present in large amounts in such places as this mine in Twin Buttes, Arizona, where blasting is dislodging copper ore. But some areas are copper-poor, and plants grown in such places suffer a deficiency of this trace element.

baby food from different parts of the United States. She found that baby foods meet, or even go over, the amount of copper that infants need daily. This was good news for mothers.

The Ten Young Men of Iran

During the late 1950s, physicians in Iran were puzzled by ten young men who came to a hospital. All were from different villages but they looked a lot alike. They were small and sexually underdeveloped for their age. Most were in their twenties, but they looked more as if they were ten or eleven. Most of them were weak and some were unable to work. The physicians examined them and found all were anemic. Every one of these men ate diets composed almost entirely of wheat bread—little meat, fish, or eggs and no fruits and vegetables. The physicians gave them iron for their anemia. The men matured physically and sexually. But the physicians were not satisfied. Growth and development were still not all they should have been. The physicians began to wonder whether the patients' diets had not

Sufficient zinc in the diet is necessary for proper growth. The three chicks on the left were small and weak, unable to stand due to abnormal leg development. They received only, left to middle, 7, 12, and 17 ppm (parts per million) of zinc in their soy protein diet. The two chicks on the right, in much better shape, were fed a different diet containing 27 ppm of zinc for the fourth chick from left and 47 ppm of zinc for the chick at the right.

HUMAN NUTRITION LABORATORY,
AGRICULTURAL RESEARCH SERVICE, USDA

made them deficient in another trace element. Could the missing trace element be zinc?

Then a research team studied some young men in Egypt. They had symptoms similar to the young men in Iran. Their diets were also made up mostly of wheat flour. When some of these patients were given a diet supplemented with iron, they improved. But the patients who received zinc actually zoomed upward, the effect was so great. Their growth was nearly twice that of the patients who got only iron supplements. One small fellow, who was 20 years old and only 39 inches tall, grew five inches in 14 months on zinc therapy. Sexual maturation among the patients getting zinc was just as great.

Actually, these findings about the importance of zinc to humans reinforced what had been found about zinc in animals some years earlier. A case of zinc deficiency was first produced experimentally in animals in 1934. Workers fed rats a controlled diet that was poor in zinc. They observed that the rats grew very poorly, lost hair and changed in hair color. Other scientists have since reported bad effects of zinc deficiency in lambs, cows, dogs, chickens, and other kinds of animals.

Over the years it has become clear that zinc works with certain enzymes. As a result, scientists have looked more closely at these enzymes. The first clue as to how zinc might act in the body came in 1939. Biochemists purified an enzyme containing zinc as a part of its structure. They found that the enzyme is vital for ridding the blood of carbon dioxide. Then, in the 1950s other enzymes were found to contain zinc, and by now 15 to 20 zinc-containing enzymes are known.

Zinc is present in some amount in all of our tissues—bone, kidney, liver, muscle, and others. But the total amount of zinc is quite small. It is about half the body's store of iron, but about 10 times that of copper. Our zinc needs and those

In an experiment conducted by Dr. James C. Smith, Jr., this rat lost hair as a result of receiving too little zinc in its diet. Other effects noted were skin irritation and poor wound healing.

VETERANS ADMINISTRATION

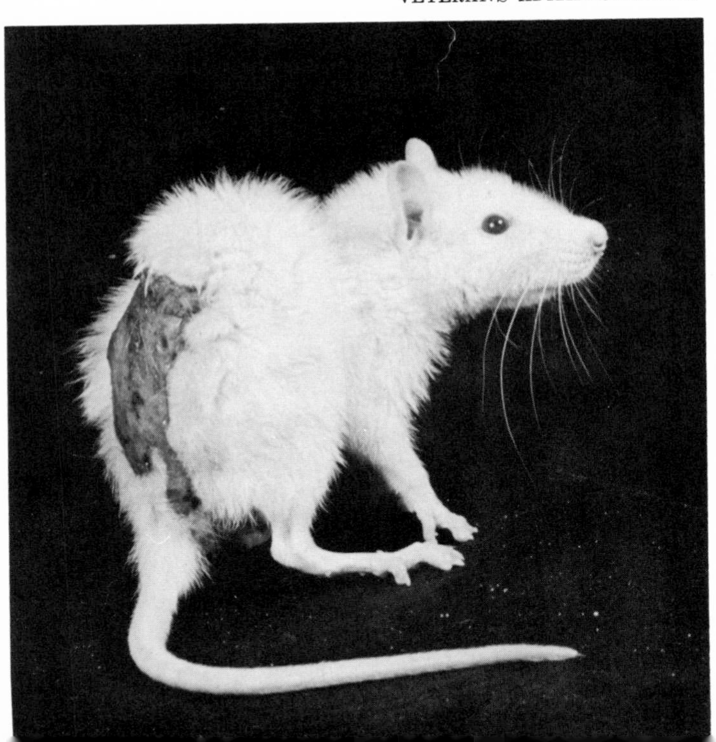

of animals must be filled through the diet. Fortunately zinc is widely distributed in foods. Most protein foods, such as milk, meat, fish, and eggs, are good sources. So are whole grain cereals and breads.

Important: Selenium and Chromium

Selenium is another middle-weight element that is valuable. Selenium made news in biological research in 1957. Before then, it was never thought to be of benefit to animals or people, since it was one of the rarer elements on our planet surface. Then, in 1957, traces of the element were found to keep the livers of rats from becoming diseased if the rats were fed diets low in vitamin E. This advance in scientific understanding soon led to practical results. A number of selenium-deficiency diseases were noted in sheep, pigs, and chickens in areas where there was little selenium in the soil. In other words, the animals were eating plants that did not contain enough selenium to meet their needs. If livestock do not get enough selenium, they may get nutritional muscular dystrophy. This is a disease in which their muscles wither and die. Selenium is now known to be vital for animals, even if they get enough vitamin E in their diets. A few cases of human selenium deficiency have also been found.

After selenium is eaten, it is carried by the

bloodstream throughout the body. From the bloodstream it moves into hair, bone, red blood cells, or other tissues. Selenium is known to make up different proteins and two muscle enzymes. It is found in the liver and kidneys more than in other organs.

How selenium works with vitamin E is not clear. The two may both help enzymes that control the amount of oxygen used by the body. The benefits of selenium have prompted farmers to ask that it be added to livestock feed, even though it is not known how much to put in. Selenium is effective in extremely small amounts, though the amount of it occurring in a certain quantity of food is hard to find out. And almost nothing is known about how much of the selenium in the diet is absorbed by the bloodstreams of animals.

People living where there are soils and plants rich in selenium may have more selenium in their blood. But most likely this is not the case, since most Americans today eat foods that come from many different parts of the country. Scientists would also like to know more about the effects of cooking on the amount of selenium in foods.

Only as recently as 1959, traces of the element chromium were found to be important to health. If animals did not get chromium in their diets, their bodies could no longer use sugar properly. The deficiency was cured by giving them chromium. Many people's bodies do not make use of sugar well as they grow older. We shall hear later about the value of giving these persons traces of chromium.

In fact, as people get older, chromium decreases in their tissues. No one is sure why. A deficiency in chromium has been linked with heart disease. Another interesting thing about chromium is that it competes with iron for a free ride through the bloodstream. The protein that chromium and iron both may attach to is called transferrin. In some persons, iron always manages to beat out chromium for transport by transferrin. As a result, these people have trouble breaking down sugar properly. The competition of chromium and iron for transport is not the only instance of trace elements' competing to take part in various biological activities.

Vanadium, a metal that is quite rare, is one of the elements found to play important trace roles in health. Four laboratories have shown that if chickens and rats do not get vanadium in their diets, they do not grow right and cannot reproduce. Like other middle-weight elements, this substance can combine fairly easily with other elements. So it probably helps animals and people by serving as an enzyme helper. Vanadium has long been known to help in various biological reactions. For example, it speeds up

HUMAN NUTRITION LABORATORY,
AGRICULTURAL RESEARCH SERVICE, USDA

Lack of enough vanadium in trace amounts makes for poorly developed feathers in chickens. The bird on the left received 30 parts of vanadium per billion parts of feed; the one on the right got two parts of vanadium per million parts of feed.

reactions that involve the catecholamines—chemicals that aid the work of the brain and central nervous system.

Fluorine has also been found, quite recently, to be of value as a trace element. Traces of fluorine—always in a compound, for this gas is extremely active—are present in widely varying amounts in animals and people. The highest concentrations of fluorine in mammals are found in bones and teeth. Fluorine may keep older women from getting osteoporosis, a disease in which the bones become less dense and more likely to fracture.

When fluorine is added to drinking water, it helps protect teeth against decay. So it has been added to the water supply in many communities. Some people, including scientists, are concerned about such additions, though. Studies have suggested that geographic, or regional, occurrences of cancer might be due to adding fluorine to the water. Certain amounts of the chemical added to the drinking water of mice also stimulated the growth of tumor transplants in them.

Scientists are also showing that fluorine is important for general growth and development. But the amount of fluorine in muscles and various organs is quite small. Its chemical properties suggest that it hooks up to the element phosphorus. It is also known to attach itself to acids in the blood. It has been shown to make tissues grow faster by activating enzymes.

Few foods contain fluorine. Seafoods have more of it than most other foods. An excellent source of the element is tea. Most water has less fluorine in it than foods do. But when fluorine is added to water, as it is in many communities, it may promote growth and development and keep bones from becoming brittle.

Nickel has been added to the list of beneficial trace elements. In 1970 researchers reported that

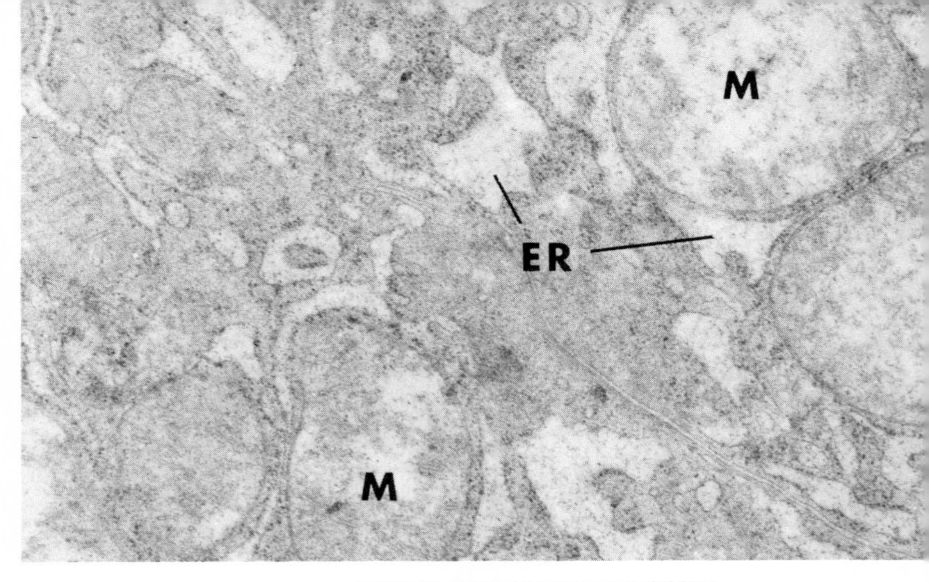

HUMAN NUTRITION LABORATORY,
AGRICULTURAL RESEARCH SERVICE, USDA

Above, a microscopic view, enlarged 24,000 times, of a chicken's liver cell. The bird was fed adequate amounts of nickel. M *shows normal mitochondria, energy-producing granules in the cell.* ER *shows normal endoplasmic reticulum, a network structure in the cell. Below we see a liver cell from a chicken that did not get enough nickel in its diet. The mitochondria and the endoplasmic reticulum are swollen, an abnormal condition.*

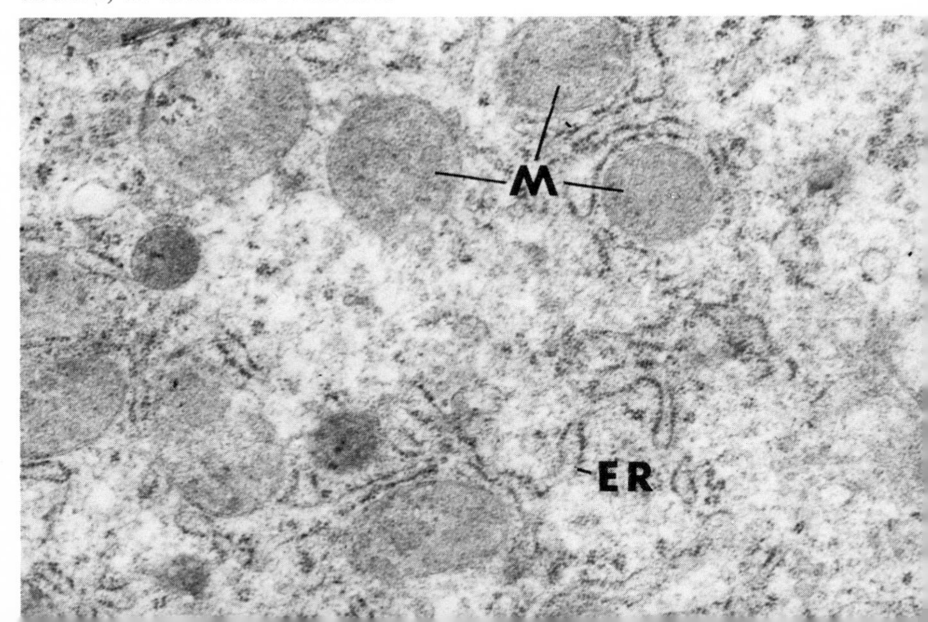

chickens kept in a special environment and fed a low-nickel diet developed thickened legs. Their legs also changed from the usual yellowish brown to a bright orange. These changes were prevented by adding nickel to the chickens' diets.

Nickel is known to spark off activity in a number of enzymes. But such activity has been shown only in experiments in pieces of tissue kept alive in special fluids. Scientists are not sure whether nickel also sets enzymes going in living animals and human beings. However, it has been detected in one of the most important compounds in living organisms. This is ribonucleic acid (RNA). This chemical is involved in carrying genetic information within cells—"instructions" for producing new, identical cells. Nickel may also be part of protein pigments that give color to tissues. Black rabbit hairs contain traces of nickel, copper, and the element cobalt. White rabbit hairs contain traces of nickel. Yellow rabbit hairs hold traces of nickel and of the elements titanium and molybdenum. The same trace elements were also present in fish hair, bird feathers, the wings of moths, and human hair.

Scientists are not yet sure what nickel does for you and me, or for the less developed animals than ourselves. One biochemist who has been working with nickel says, however, that nickel is important to biological activities in the liver. He thinks that it may act on the surfaces of cells, or perhaps on deoxyribonucleic acid (DNA). DNA contains the genetic information that RNA passes on.

Until nickel's role in animals and people is better understood, it is hard to say how much of this metal we should get in our foods every day. But nickel is present in many things we eat, particularly in whole-grain breads and flours and cereals. There are fairly large amounts of nickel in vegetables, tea, and cocoa.

6/People Exploit the Elements

As we have just seen, there seems to be a need for trace elements among all living things. But the kind and amount of trace elements taken in varies from one organism to another. Intake is decided by the particular needs of a plant, animal, or person; it is also decided by the amount of the elements in the environment. Plants, animals, and humankind have probably evolved in such a way that they could take best advantage of the amounts of trace elements around them. For millions of years an ever-changing but delicate balance has probably existed between life's need for trace elements and the amounts of trace elements present that life can use.

Then men and women appeared on earth. They had superior intellects and were able to change their environment. The natural balance of trace elements in the earth's crust, air, and waters changed. If our ancestors had not used the elements, we would not have reached the level of civilization we now enjoy. But at the same time, people's use of them has upset the natural balance of trace elements.

Some of these have become too scarce for the needs of living things; others have become too plentiful. There is especially a danger in an overabundance of trace elements—in short, trace-element pollutants. These pollutants, as we shall see, are widespread. Some of them are hurting our health and well-being and damaging many kinds of plants and animals.

When did it all start, people's use of the elements? Archaeologists, or scientists who study the remains of past civilizations, have found weapons and ornaments used by peoples of 100,000 to 35,000 years ago. Many of these objects contain metals. Metals are a large group of middle-weight and heavy elements. Some of them have been discussed already.

Gold and Silver

Gold was probably the first metal to be used by primitive peoples. Although it was scarce in the earth's rocks and soil, it had many fine qualities. It was resistant to rust. It was soft and could be hammered at air temperature into different shapes. Best of all, it was glittering and beautiful. So it was hammered into crowns, necklaces, bracelets, and urns. Gold was much wanted by the wealthy and powerful. It belonged to kings, queens, sultans, and pharaohs. King Solomon owned gold; so did King Midas. And the Queen of Sheba.

Silver was probably the next element to be used by ancient peoples. While not as lustrous as gold, it too was beautiful. It was also less scarce in the earth's crust. So it too was taken from the earth and hammered into plates, goblets, bracelets, armlets, fans, and jugs.

But gold and silver were too soft to be made into weapons. Early peoples, however, came upon a reddish metallic element, copper. This was stronger than gold and silver. It too could be beaten into different shapes at air temperature. Copper was the first metal to be extracted from its ores—that is, from copper mixed with compounds and other elements in the earth. The copper was heated into a liquid-like state so that the other elements separated out. The process of separating an element from its ore is called smelting. Smelted copper was then hardened and hammered into cooking utensils and weapons. The earliest known extractions of copper ores date back to 6000 B.C.

Then our ancestors discovered a bluish-white metal that melted easily out of rocks. It was tin. At first they were probably disappointed

Metals have been mined as far back in human history as 4000 B.C. *This wood engraving, showing combustion in a mine, is from* De re metallica, *a pioneering work on minerals and mining, by Georg Agricola, a German physician and scientist of the sixteenth century.*

because it was too soft to make into utensils and weapons. Over the years they found they could melt tin, together with copper, into a hard metal mixture, or alloy: bronze. This substance was to have an enormous impact on civilization. Now humanity had a metal it could turn into knives, axes, cooking pots, utensils, axles for cart and chariot wheels, swords, and armor.

There is archeological evidence that iron, a silvery-white metallic element, was used by people as far back as 4000 B.C. The iron was probably taken from meteorites from outer space that hit earth. A good reason for believing this is that tools found from that period contain a lot of the metallic element nickel. Nickel is mixed with the iron from meteorites. Iron found naturally in the earth's crust does not contain any. The meteoric iron could probably be hammered at air temperatures in some cases, as could gold and silver, or at most in the heat of a wood fire. But the iron became harder than either gold or silver.

No one knows when iron was first smelted, but we have a rough idea. An obscure group of people known as the Hittites lived in the mountains near the Black Sea in 1200 B.C. They were probably the first people to smelt the metal. As findings from Palestine show, smelted iron was made into knives, hoes, sickles, and other agricultural tools. In Greece in 500 B.C. it was made into arrows, chain links, files, chisels, swords, razors.

Tools and weapons dating back 3000 years include some of the bluish-white metal lead, which may have been in the materials accidentally. Or it may have been intentionally used. Not until 500 B.C., however, was lead used extensively in this way.

From 1600 to 600 B.C., peoples in Palestine, China, and Persia fused the metal zinc with copper. The resulting alloy was brass. The first real impact brass had came during the role of the Roman empire. The Romans made coins of brass. In fact, they exploited many elements during their rule. They mined vast amounts of silver, copper, lead, and tin in Spain. Tin mines in Cornwall, England, came under Roman rule. They beat lead into sheets and pipes and were the first to use tin for lining food containers.

By the sixteenth century, the smelting of copper, lead, tin, and iron was carried out in small blast furnaces. By the seventeenth century, the American iron industry had made its start. The Indians of North America learned from the French pioneers how to smelt lead ores. Before, the Indians had used gold and copper, but they did not dig the ores of these elements or smelt them.

During the 1700s, one product became available that was to have more of an impact on civilization than any other. This was steel. It was smelted from pig iron, which was iron mixed with carbon, sulfur, manganese, silicon, and phosphorus. The first good process for making steel was invented in 1740. The process, and subsequent improvements on it, prepared western Europe and the United States for the Industrial Revolution.

From Windmill to Steam Engine

The Industrial Revolution took place over the eighteenth and nineteenth centuries. Serfs

The Industrial Revolution called for vast quantities of coal to keep it going. Not only adults but even children were made virtual slaves doing back-breaking labor in British coal mines during much of the nineteenth century. This picture is from the Report of the Children's Employment Commission, on Mines and Collieries, *published in London in 1842.*

became farmers or migrated to growing factory towns. In towns, handicrafts were replaced by machines and large factories. Not just the rich but the poor as well could sometimes makes more money than they had before if they were clever and hard workers. It was an age of great technical discoveries. More rapid ways of making textiles appeared. The windmill gave way to the water wheel and steam engine. These provided power for factories. Iron and steel replaced wood in the construction of harbors, bridges, and railroad tracks. Steel went into the construction of sewing machines, cranes, elevators, hydraulic pumps, carriages, fire engines, bicycles, sugar and salt pans, and many other items.

By the middle of the nineteenth century, the processing of steel became largely what it is today. Air was blown through melted pig iron to remove unwanted elements. Refinements have been added to the process since then, of course. Other elements are added to steel to make it more useful for various purposes. Nickel, molybdenum, and vanadium are often added to steel that is used in automobiles and airplanes.

Even though steel became king of the industrial world, other elements continued to be exploited. Thousands of Americans rushed to California in search of gold, and to Colorado in search of silver. If you ever go to Leadville, an old mining town southwest of Denver, you will hear how silver mining was a rich and colorful part of the settling of the West.

At the height of the silver boom, Leadville was a wild frontier town, where law and order were the responsibility of a gun-fighter who was appointed city marshal. The town's most famous resident was H. A. W. Tabor. He arrived in Colorado in 1859 and later bought the fabulous Matchless Mine in Leadville. But when silver prices collapsed he was ruined. On his deathbead he told his wife, "Hold onto the Matchless." She did. She lived in poverty in a cabin by the mine until she died in 1935. She still hoped—in vain—that the mine would again produce a fortune in silver.

A Flood of Elements

Aluminum is one of the most abundant elements in the earth's crust. It was first isolated in 1825. Unlike other common industrial metals, such as iron, copper, zinc, and lead, aluminum is not produced by the direct smelting of its ores. Electricity is used instead to separate aluminum from its ore. The modern electrolytic method for producing aluminum was discovered in 1886. During an exposition in 1900, many aluminum parts for automobiles were displayed. An aluminum anchor used for several balloon trips was also shown. It was used in the framework of airships in World War I. After World War II, buses and trucks got aluminum bodies. High-speed trains were made of aluminum alloys. Today the element is turned into keys, thimbles, brushes, combs, opera glasses, knives, cooking utensils, windows, doors, roofing, store fronts, canteens. Aluminum foil is wrapped around chocolate bars and cheeses, and is used to wrap leftover foods and broil meats. It is also used in radios and radar equipment. Aluminum is often added to steel, and for many purposes.

Silvery-white calcium is the fifth most abundant element in the earth's crust. It is found in rocks and soils, and mixed with other elements.

REYNOLDS METALS COMPANY

Aluminum, one of the most abundant elements, being prepared in a reduction plant. The mechanical crust breaker in the foreground stirs up the contents of a reduction cell. The mixture is drawn from the cells by the large rolling crucible, or melting pot, on wheels that rolls after it.

In 1808 a scientist showed that lime was a compound of oxygen and another element. He named the element calcium. Other scientists learned to use electricity to separate calcium from its salts. Practically all production of calcium today is by electrolysis (producing chemical changes by passing an electric current through a solution). Although the entire production of calcium today is much smaller than that for aluminum, calcium still has a number of uses. It is used to strengthen industrial alloys of aluminum, copper, lead, magnesium, iron, chromium, nickel, even steel. It is used in radio and television parts.

The bluish-white element chromium was first isolated in the nineteenth century from its ore, which includes iron and oxygen. During the early twentieth century, chromium became important to steelmaking. Today chromium is used to improve the strength of not just steel but also of nickel and other metals. It is used in the production of pigments that in turn are used to color cements, plastics, woods, and metals. Chrome oxide green is the most stable green pigment known. Chromium yellow is an excellent paint for wood and metal. Chromium is also used to make prints on textiles and to harden photographic films.

The blackish-gray, nonmetallic element iodine was discovered during the nineteenth century as a result of war research. During the Napoleonic wars, France was isolated from other countries by the British Navy. The French could not get outside souces of the chemical compound potassium nitrate for gunpowder. So they used ashes of seaweed to convert calcium nitrate into potassium nitrate. During the process they discovered that seaweed contained an unknown element. It turned out to be iodine. The harvesting of iodine from seaweed subsequently became an important business along the coasts of France and Scotland.

Today iodine is used in making photographic film. People are given injections of it so that the internal structures of their bodies show up in X-ray photos. You may apply iodine solution to cuts or scratches on your body to keep them from becoming infected.

On the whole, elements that have been used by people for several, or many, centuries continue to be used in the modern world. Consider the nonmetallic sulfur, arsenic (which is somewhat like a metal), or the precious gold and silver.

Sulfur is present in volcanoes and spouting geysers. It has a nasty smell. In ancient times

Geysers, like this one—Old Faithful in Yellowstone National Park—are among the natural features that add sulfur to the environment. It is sprayed up in the water, along with other trace elements.

UNION PACIFIC RAILROAD

witch doctors burned sulfur to drive away evil spirits. People discovered that its fumes can kill insects. Sulfur was burned to cure ills. The Romans used it in weapons that carried flames to the enemy, and it was used in gunpowder in medieval times. During the eighteenth century its commercial value came to light. Today sulfuric acid is the cheapest acid available to industry. Sulfur goes into fertilizers, newsprint, wrapping papers, rayons, steel, paint pigments, pesticides.

Practically all arsenic is produced as a by-product from roasting the ores of copper, lead, and others. For a long time arsenic was "the" poison to do someone in with. Being in a white powder form, it could be slipped into a glass of wine without someone seeing it. Then during the nineteenth century arsenic turned commercial; it was put into wallpapers and artificial plants. Today it is used in insecticides and weed-killers. It is also used as a decolorizer in the glass industry, and to preserve wood.

Gold continues to be the most prized of all the elements. It is made into watches, earrings, bracelets and other jewelry, and is used to fill teeth. As gold supplies become more and more scarce, gold is being hoarded by many people. Although countries' monetary values are no longer placed on how much gold they have, it is still a symbol of wealth. Silver also continues to be in high demand; it is used in silverware, jewelry, tea and coffee sets, and in the processing of foods and beverages.

7/ *Trace Pollutants in Air, Soil, and Water*

In a mountainous area of Yugoslavia, not far from the Austrian border, there is a lead mine and smelter. The mine has been producing rich lead ores for 300 years. Until 1969, 300 tons of lead were discharged into the air each year from two smokestacks on the smelter. Then filters were put on the stacks to keep lead from escaping. However, a chemist suspected that lead was still escaping in spite of the filters. He set out to see whether he was right.

He collected samples of lead in new-fallen snow in different directions and at increasing distances from the smelter. He gathered all of his samples in the spring of 1970. He put each sample of snow and lead in a cylindrical tube. As soon as he did this, he closed the tubes at once. In this way lead in the snow samples could not escape, and traces of lead from other sources could not contaminate the samples. As he suspected, lead was still escaping into the air. Snow closer to the smelter contained more traces of lead pollution than did snow farther from it.

This experiment showed what other experiments have since confirmed: people's exploitation of lead is upsetting the natural balance of lead in the environment.

Actually, lead pollution is nothing new. As we have mentioned, lead smelters have been operating for 400 years at least. In fact, traces of many elements have probably been pouring from man-made sources into the air, water, and soil as long as lead has. What is new is that chemists can now detect trace-element pollutants. They had to wait for sensitive scientific instruments to be invented before they could do this.

Mines and Smelters: Mixed Blessings

Let's take a look at what chemists now know about the sources and distribution of various trace-element pollutants.

As the Yugoslavian experiment showed, smelters are a major source of lead and other pollutants. For example, cadmium is a relatively rare metal in the earth's crust. Yet it is rich in zinc, lead, and copper ores. When these ores are smelted, cadmium can escape as pollution.

Mines too can cause pollution. Virtually all arsenic produced for industrial purposes is re-

DARTMOUTH MEDICAL SCHOOL

Dr. Henry A. Schroeder, Professor of Physiology Emeritus of Dartmouth Medical School, at work in his trace-elements laboratory in Brattleboro, Vermont. "Cadmium and lead are harming people right now," says Dr. Schroeder.

covered as a byproduct of lead, copper, and gold mining. One gold mine in the West of the United States produces 14,690 tons of arsenic each year—actually more than is needed. Much of the left-over arsenic gets into the environment. Mercury and cadmium can also escape from mines.

Another major source is industrial processing of elements. Several chemical companies decided to look into this. One firm found that its processes give off 300 different kinds of pollutants!

In 1970, 4800 pounds of lead were discharged every day into the Mississippi River between Baton Rouge and New Orleans. In the same area some hundred pounds of arsenic were also being discharged every day. The industries that were pouring out these pollutants were soon tracked down. One of the main sources of arsenic pollution is factories that make fertilizers, detergents, and pesticides.

Beryllium pollution may result from producing coal or rocket fuels. Traces of boron are given off in making glass and medicines. Vanadium is given off in the making of steel alloys.

Cadmium is a major trace pollutant. One source is processing zinc, lead, and copper ores. Another is the electroplating industry, which adds cadmium to other metals to make them resistant to corrosion. Making television picture tubes, fungicides, and shields for nuclear reactors adds still more.

The burning of gasoline and other fuels appears to add trace-element pollution. For example, 1600 pounds a day of lead, 7900 pounds a day of zinc, 5000 pounds a day of cadmium and 300 pounds a day of chromium were discharged into the Houston channel by ships in 1969. This was vastly more than nature puts into the water. There was 63,000 times more lead, 108,000 times more chromium, and 15 times more cadmium than is usually present.

Traces of boron are given off in the burning of coal, and in the use of cleaning agents. Selenium pops out when coal and sulfur are burned.

AGRICULTURAL RESEARCH SERVICE, USDA

In checking for the effects of chromium in our food, chromium-containing compounds are extracted from different foods, above. After they are incubated with the fat tissue of rats, their effect on the oxidation of sugar is measured, with the help of radioisotopes. After the radioactivity is measured, the pieces of fat tissue, below, are weighed and thus the strength of the test compound is found.

U. S. GEOLOGICAL SURVEY

This historic photo of Nob Hill, San Francisco, after the famous 1906 earthquake, shows one of many ways that trace elements can enter the environment—from natural phenomena such as earthquakes, volcanoes, geysers, erosion, fires, floods, and the weathering of paints and many other substances.

Cadmium enters the environment when people smoke cigarettes and pipes. Traces of vanadium, bismuth, tin, and zirconium escape into the environment when coal and petroleum are burned. A million pounds of mercury slip into the American environment each year from the burning of coal, and more enters from paints on buildings and from burning electrical batteries in incinerators.

Two to three times as much lead enters the environment from dumping, weathering, and burning of paint pigments and metal products as comes from the use of fuel. The burning of plastic products, from bottles to baby pants, adds traces of cadmium to the air. Cadmium also escapes in trace amounts when fertilizers, fungicides, motor oils, and rubber tires are used. One chemist found arsenic in the Kansas River, apparently from the dumping of detergents into the river. Detergents also release traces of phosphorus.

Many pollutants end up in the air. Lead is present in the air of some cities in high quantities. Chemists in Sweden report that most air contains only a little cadmium but that air near certain factories has a lot more of it. Pollutants also find their way into oceans, rivers, streams, even into drinking water. Arsenic in fertilizers and detergents may end up in both city and country rivers and streams. In a study completed in

1971, the U. S. Geological Survey found that many of the 720 rivers and streams they sampled contained high levels of cadmium. A few waterways had high levels of lead and mercury.

Mercury finds its way into the bottoms of lakes. Much of the mercury pollution in the Great Lakes became apparent in the spring of 1970; it was traced to the dumping of large amounts of industrial wastes. This discovery came as a surprise to many people. But there was every reason to expect such pollution, since Sweden and Japan had already found large amounts of this element in their waters. Much mercury has been found also in Lake Delta, New York. An extensive study of trace pollutants in Lake Michigan showed that 2700 tons of copper and 760 tons of nickel are poured into the lake each year, probably by industrial plants.

Nickel ore being hauled in a modern nickel mine. Lake Michigan receives hundreds of tons of nickel every year as a result of industrial activities. Rivers and lakes are particular sufferers from pollution of many types.

THE INTERNATIONAL NICKEL COMPANY OF CANADA, LTD.

Trace pollutants in drinking water are tricky to pinpoint. Although chlorine and aluminum may be put in drinking water at the water-treatment plant, these trace elements may not still be in water that you and I drink. Somehow or other they escape on their way to our faucets. On the other hand, drinking water can pick up traces of cadmium, copper, nickel, or lead as it passes through pipes into homes and offices.

Other trace pollutants mix with soil. Mud from the Houston channel showed large amounts of cadmium, lead, and tin, probably from the burning of fuels by ships in the channel.

Grass sampled alongside highways contained a hundred time more lead than grass not exposed to automobile exhaust containing lead.

When these young heifers grow up and start producing milk, it will be milk containing lead and other pollutants, taken in with the grass they eat. Pastures near much-traveled highways receive more lead from car exhaust than those far away from heavy traffic.

USDA

Large amounts of nickel, cadmium, and zinc have also been found alongside highways. Apparently these come from automobiles. As mentioned earlier, cadmium is present in motor oils and automobile tires and could escape as automobiles whiz along highways. Zinc and nickel are also found in motor oils.

There is more lead in the soils of marshes if the marshes are in or near cities or suburbs. A lot of lead is present in street dust, and in the soil of city parks.

Pollutants are often concentrated over large cities, where they are mostly emitted. Large American cities have 20 times more lead in their atmospheres than do rural areas of the United States. Larger American cities have 2000 times more lead in their atmospheres than does air over the Pacific Ocean. But trace pollutants can be scattered by winds and moved far from the places they come from.

Larger particles of pollutants usually don't travel very far through the air. But very small particles are often raised by winds to higher altitudes. As a result, they can scatter over vast distances. If trace pollutants are blown high enough in the sky, they can sail clear around the world—from Louisville, Kentucky, to Peking, China, for example.

Antimony escapes into the environment as a byproduct of steel production. Traces of antimony were found in excessive amounts in Canada, Bermuda, and the Bahamas. This substance could not be traced directly to power plants or to other sources of pollution, so it looked as if it had been carried to remote areas from far-off industrial sources.

One chemist wondered whether people's use of the elements might have affected one of the most remote parts of the earth, the South Pole. So he and his team went there to find out. Problems in collecting trace elements in this frigid, snowy land proved immense. The cold was so great that some of their collection equipment broke in their hands. Still, after three years of perseverance, they finally got the samples of ice and snow they wanted. Then they brought the samples home and analyzed them for trace pollutants.

The Cleanest Place in the World

"The South Pole," they reported in 1973, "is still the cleanest place by far on earth." Vanadium at the South Pole was a million times less prevalent than in Boston, where power plants and cars give off very much of it. The South Pole was even ten to 50 times cleaner than Hawaii, which is largely free of pollutants.

Even so, the chemists did find a little "chemical garbage" at the South Pole—notably manganese, selenium, iron, and cobalt. Such findings were remarkable because the samples were taken from ice and snow that were 500 miles from land. So industrial pollutants had to travel at least 500 miles, if not more, to show up there.

It becomes obvious, then, that the enormous use of chemicals is flooding the whole world with traces of elements. Icy wastelands, craggy peaks, even your treehouse or favorite hideout are not being spared from this pollution.

Just being able to detect trace pollutants, however, is not enough. What is happening raises other tough questions. These are questions that chemists have not answered very well, or questions they would like to answer better. For example, what chemical and physical changes do trace pollutants undergo after they enter the environment?

In 1973, a chemist followed the changes pollutants underwent as they moved down the Amazon River in Brazil and down the Yukon River in Alaska. All of the trace elements he studied changed physically or chemically. Some traveled by themselves; others attached themselves to water molecules. Some formed crystals; others did not. Most remarkable, identical trace pollutants experienced similar changes in both the Amazon and the Yukon, in spite of one river snaking its way through steaming jungles and the other coursing its way into the icy Bering

Sea. But the Amazon and Yukon Rivers are not very polluted.

Chemical Detectives

Another mystery chemists would like to unravel is this: Which trace elements are really strangers in a region and which are naturally present there? But an answer is hard to come by, because many trace elements are often mixed together in one geographic area.

For example, one group tried to pin down the origin of trace elements in and around Lake Windermere in Britain. From 1970 to 1972 they took weekly or monthly samples. They found that some 30 trace elements were regularly present; another nine popped up in the samples from time to time. The only way they decided whether the trace elements were really pollutants was to make some educated guesses.

Because there was more chlorine in the lake than is naturally present in oceans, they decided that it should be considered a pollutant. Also, more chlorine appeared in the air over Lake Windermere and in the lake itself during winter than it did at other times of the year. The chemists concluded that it probably came from people burning coal during the cold months.

They found an increase in traces of vanadium too during the winter and decided it came

THE ANACONDA COMPANY

Casting bars of copper for use in making electrical wire. Copper is one of many "problem elements"—wonderfully useful to the human race but potential sources of trouble if too much of them spreads through our environment.

from the burning of heating oil. Traces of lead, zinc, copper, rubidium, calcium, and bromine, they concluded, came from factories.

Another question the chemical detectives would like to answer is: How much of this pollution observed today is due to what is going on right now, and how much is due to a build-up through several centuries—in fact, since people first started using the elements? Chemists argue, and have no final answer to this question. Some are convinced that trace pollutants have been building up in the environment for not only centuries but even thousands of years. Fossils of ancient fish have been found to contain large amounts of mercury. One scientist tested 17 samples of fish from archeological sites in the United States and Peru; he found mercury in 12 of the samples. Two of the fish that contained mercury lived 750 years ago. And the levels of the metal found in these fish were the highest that have ever been detected in marine fish, even those living today. This unexpected result suggests that mercury has been in the oceans in large amounts for hundreds of years.

Some studies suggest that the surface waters of oceans today have ten times more lead in them than when people first appeared on earth, about two million years ago. As a result, some chemists estimate that people have been polluting the air, soil, and water with lead for at least 5,000 years. After all, lead pollution could have started that long ago. As we have seen, tools and weapons dating back 3000 years have been found to contain lead.

Other chemists admit that trace-element pollution may have started a long time ago but insist that most pollution has taken place in recent years. For example, careful dating methods showed that the icy waters of Greenland contain lead that dates back to 800 B.C. Yet most of the lead taken from these waters was carried into them during the past several hundred years. Tests made in the air over Greenland indicated that lead pollution has increased dramatically since the Industrial Revolution. It was shown to have increased 16 times over since 1900.

Still other investigators believe that increases like this are almost entirely the product of the twentieth century. One chemist who has spent much of her life studying trace elements says: "It has been only during the past hundred years that people have been polluting the environment with large amounts of elements. The pollution has come mostly from the smelting of elements, from industrial applications of elements and from the burning of fuels."

In one study, she set out to find out how much lead was present in different levels of marsh soil. Soil closest to the surface was only a couple of years old. Soil a few feet down was several decades old. The amount of lead in each level of soil would indicate how much lead pollution took place during a particular era.

She dug a large hole in the soil. She dug it deeper and deeper. Finally the hole was so deep she was down in the muck, shoveling soil up and over her head. Finally she dug herself deep enough to get soil samples from the past 50 years. She found that dirt from the past decade contained far more lead than did that of 50 years ago.

One of the most crucial questions chemists want to answer is whether trace element pollution is on the increase. Preliminary studies suggest that it is.

In 1968, lead in the air over San Diego was reported to be increasing 5 per cent each year. In

1971, the Environmental Protection Agency reported that lead is also increasing in the air over other large cities. The EPA took measurements in Cincinnati, Los Angeles, and Philadelphia in 1961, and again in 1969. They showed that lead had increased 16 per cent in Cincinnati, 50 per cent in Los Angeles, and 18 per cent in Philadelphia.

In 1970, 27 per cent of America's streams and shores were polluted with trace elements. In 1971 it was 29 per cent. One reason for the increase was that traces of nitrogen and phosphorus in fertilizers were running off farm soils and into nearby streams.

When forests are cleared, trace elements in their soils are flushed into streams by the rain. As a result, these chemicals that once helped plants grow now became a nuisance. Clearing of forests was widespread after the Civil War. It is even more common today.

With increasing demands for fuel and other substances locked in the earth, there is a step-up in surface mining of copper, zinc, silver, lead, sulfur, calcium, and other elements. As the earth's surface is ravaged by mighty bulldozers, bits of the mined elements escape into streams and air.

Chemists can now detect trace pollutants in parts so small they cannot be seen with even the most powerful microscope in the world. Such small amounts are almost beyond the wildest human imagination. During the next few years, trace-element detectors will become still more sensitive; even smaller particles will be detected. But even using such elegant methods, it will probably be many years before chemists can tell us eveything we want to know about where this dangerous chemical trash comes from and what we can do about it.

8/ What Are Trace Pollutants Doing to Us?

During the past few years, chemists have looked with an eagle eye at the way trace pollutants are distributed in the air, soil, and water. At the same time, other scientists—such as biochemists, physiologists, and toxicologists—have been studying the effects of these pollutants on living things. They are making some vital discoveries that affect you and me.

One of these is good news. It is that trace pollutants may help plants, animals, and people. Traces of zinc, copper, chromium, vanadium, and some other elements that are excessive in air, soil, and water may actually be beneficial to our health.

Another discovery they are making is not exactly good news or bad news. It is that many trace pollutants get into the tissues and cells of living things but apparently do not harm them.

Some 30 such substances have been identified. They include such unlikely elements as aluminum, antimony, silver, and gold. You may have traces of shiny aluminum or of glistening gold and silver in your body right now. But they are probably not hurting you, if only because they are present in such tiny amounts.

Still another finding of these scientists is definitely bad news. Some pollutants are hurting life. Or they have the ability to hurt life if we let them go unchecked.

Figuring out which trace pollutants are bad isn't easy. Even the most helpful of them can be harmful if we take them in too large amounts. Iron is valuable to plants, animals, and to you and me in specific small quantities. But miners who dig iron ores are exposed to somewhat larger amounts of iron and over a long period of time. As a result some miners get cancer. Zinc is valuable to plants, animals, and people in certain trace amounts. But plants growing near zinc smelters are in danger because they can get high amounts of the element from these smelters.

To Each Its Own Reaction

Whether a trace pollutant is harmful depends on the reaction of a particular plant, animal, or person to it. Plants may respond differently from animals to the same amount of a trace

OAK RIDGE OPERATIONS OFFICE, AEC

Plants often react differently from animals to various amounts of trace elements. Here strontium is being applied to the leaves of tomato plants.

pollutant. Animals may react differently than people do. An apple tree may be hurt more than a grape orchard. A fox may be hurt less than a wolf. You may be hurt less than I am. There is, unfortunately, no easy rule of thumb about which trace pollutants are the bad guys and which are not.

In addition, these substances may not be harmful when they leave the mine, factory, incinerator, or other source of human exploitation of the elements. But they may become dangerous when bacteria in the soil change them into different chemical forms. Bacteria can transform traces of harmless industrial mercury so that it becomes deadly for plants and animal organisms, including ourselves.

Often pollutants hurt life slowly, their final effect not being all that apparent. Plants may only wilt at first. People may feel tired, nauseated, dizzy. Often the symptoms of early trace-pollution poisoning are confused with symptoms created by other diseases. For example, children who suffer mildly from lead poisoning may be tired and grumpy. But their parents and doctors do not suspect the real cause; after all, we are all tired and irritable at some time or another. If trace pollutants accumulate enough in an organism, symptoms then become obvious indeed. Animals and people may go into convulsions, then into a coma—a very deep unconsciousness—and then die.

Once harmful trace pollutants enter a living organism, they combine with particular tissues. In animals and people, nickel goes into the lungs. Cadmium attacks the kidneys. Mercury has a "preference" for the brain and spinal cord. Lead settles mostly in bones and teeth. But it is at the cellular level—the operation of each cell acting by itself in addition to being a member of a team—that trace pollutants do their ultimate damage.

One of the most common tricks such pollutants play in cells is to replace trace elements that help enzymes fulfill various activities. This replacement is wrong from an enzyme's point of view. The trace element it needs hooks up to it in a nice, snug fit, just the way the right key fits correctly into a lock. But an atom of the trace pollutant that replaces the desired trace element is a different shape and has different electrical qualities. When it hooks up to the enzyme, it doesn't fit nicely at all. It is like a wrong key jammed into a lock. When a pollutant jams an enzyme, the enzyme can no longer carry out its usual activities.

Heavier trace pollutants are often enzyme-blockers. This is because many middle-weight trace elements are vital to enzymes, and heavier trace elements have a tendency to replace light trace elements they resemble in atomic makeup. Heavier trace pollutants are not the only culprits, though. Even middle-weight pollutants can block enzymes if they are present in an organism in large enough amounts. And whether a trace pollutant is harmful or not is also partly determined by its atomic nature. For some reason that scientists have yet to determine, beryllium is one of the most poisonous of all trace elements. Yet it is one of the lightest elements.

Let's see which trace pollutants *are* hurting plants, animals, and people, and which *may possibly* do harm. As we'll soon see, there are some surprises. A number of things play a part—the nature of an element, its presence in the environment, and the fact that certain organisms respond badly to certain elements and not to others.

9/ Trace Pollutants Hurt!

Several thousand years ago, the Romans were the most powerful people in the world. They spread their empire from Italy through the Middle East, northern Africa and Europe. But after several centuries, the men and women who built the empire lost their enthusiasm, creativity, ability to rule. They found it difficult to have children. Without successors, their empire crumbled. One of the greatest eras in the civilization of the world came to an end.

Today, a number of scientists believe that lead pollutants had a lot to do with the downfall of the Romans. The Roman rulers ate and drank from vessels that had been made from lead. As a result, traces of lead could have gotten into their foods and wines and could have slowly poisoned their bodies. Traces of lead could have made them feeble-minded, listless. Lead poisoning could have made it difficult for them to reproduce.

The Tragedies of Lead

Whether lead pollution and lead poisoning partially or fully explains the fall of the Roman Empire is not known. But one thing is sure: lead pollution and lead poisoning did not stop with the Roman Empire. Hundreds of years later, some of the pioneers who settled the United States were poisoned by drinking rum from lead containers. Generations of American children and of European children played with toy soldiers made of lead. Those children who nibbled on the soldiers sometimes became sick or even died. During the early years of the twentieth century, lead poisoning was common among house painters, since lead was common in paints (and still is found in some).

Today, lead pollution is even more of a health problem. In fact, considering that so much lead is in the environment, and it is so poisonous, this heavy element is probably the leading trace-pollution danger. Today, Americans are estimated to have a hundred times more lead in their bodies than people who lived several thousand years ago.

People living in and near cities are usually more exposed to lead pollution than are people living in the country. There is 20 times more lead in city air than in country air, and 20,000 times more lead in city air than in air over the ocean. This is because big cities are the major sources of industrial lead pollution, pollution from the automobile exhaust of gasolines with lead added to them, the dumping of used lead, and the weathering of paint.

In one scientific study, a great deal more lead was found in people living outside Los Angeles than along the coast of California. In another study, policemen in Philadelphia were found to have lots of lead in their bodies; people living in downtown Philadelphia somewhat less; those commuting to Philadelphia to work, still less; and people living in the suburbs less yet. A Swedish study showed that there were high percentages of lead in people living in the city but not so much in farmers living far from the city. Some 15 times more lead was found in people living in Ann Arbor, Michigan, than on an Indian reservation far from the sources of lead pollution.

Children are especially open to lead pollutants. For one thing, they are closer to the ground, and thus to automobile exhaust, than adults are. One scientist went around dropping lollipops on streets, park benches, rooftops, and other places. He found that lollipops dropped on or near the ground picked up more traces of lead than did lollipops dropped farther from the ground. So even if you are careful not to eat lollipops that have fallen on the ground, you may be exposed to more lead pollutants than are your older brother or sister, or your mother or father.

Many children living in older tenement houses in cities are tempted to nibble on paint that has flaked from the walls. This paint contains large amounts of lead, which are easily absorbed by the body.

The amount of lead a youngster has in his or her body may show up in baby teeth. A dentist got 69 baby teeth from dental clinics serving city children and 40 baby teeth from dentists serving suburban children. These results again show that people living in cities—particularly children—are more open to lead exposure than people who live outside of cities.

There are complications in the picture, though. Alaskan Eskimos were found to have three times more lead in their bodies than Americans living in Nebraska, Iowa, and other midwestern states. How could this be, since the Eskimos live far from cities? Scientists think it may be because Eskimos are not exposed to as much sunlight as Americans in the Midwest are.

NATIONAL PAINT AND COATINGS ASSOCIATION

Old, flaking lead-base paints are a source of danger, especially to young children. Keeping buildings in good condition and using nonlead paints are ways of lessening this danger.

In one experiment, mice were exposed to both lead and sunlight, and other mice were exposed to lead only. The mice in the first group got less lead in their bodies than did mice in the second group. These results imply that sunlight can help keep lead out of our bodies. So even though Eskimos may not be exposed to as much lead as are people in the Midwest, they are more vulnerable because there is virtually no sunlight in the North Pole region during many months of the year.

Lungs, Stomach, and Toothbrush

Air is a major source by which people take in lead pollutants. A third of the lead that gets into people living or working in cities is breathed from the air. But the other two-thirds comes from foods, beverages, and drinking water. Scientists estimate that of the lead we eat, 39 per cent is in meats, including chicken; 22 per cent in vegetables; 17 per cent in bakery products; 10 per cent in juices; 9 per cent in dairy products; 3 per cent in other foods.

You can also take in lead by brushing your teeth. A dentist tested six brands of toothpaste, since the tubes are made partly of lead. He sampled toothpaste from deep inside full tubes and from tubes that were about empty. He found that all of the samples contained lead, but especially samples from the near-empty tubes. He figures that if children brush their teeth twice a day, they get more lead into their bodies through toothpaste than they do from food. This is bad news if you want to keep your teeth clean. (Another bit of bad news is that practically all toothpastes contain sugar, which is a principal cause of tooth decay. Perhaps we should think of this sugar as a kind of pollutant too.)

Like people, the less developed animals and the plants take in more lead if they are in or near cities. A lot of lead was found in a muskrat taken from a marsh near a Connecticut city. Large amounts of lead were found in insects in a city back yard, but less was found in insects in the Green Mountains of Vermont. More was found in oak, hazel, and ash trees in London than in the same kinds of trees in the English countryside. In one study, grass near the roadside was found to have four times more lead in it than grass grown 500 feet from the road. In another study, tomato, cauliflower, rye, and potato plants grown near highways had more lead in them than plants grown farther away. The more lead pollutants plants are exposed to, the more lead they take in through their leaves, and probably their roots too. Rough, hairy leaves take in more than smooth leaves do.

What happens when lead gets into people, plants, or animals? The tiny traces build up. If they build up enough, they can seriously damage health and even kill them.

No other trace pollutant has accumulated in people to levels so close to those causing obvious poisoning. Over a fourth of all American children, 5 per cent of the men, and 2 per cent of the women are estimated to have levels of lead in their bodies that border on toxicity, or ability to poison. An American scientist who has pioneered in the study of trace pollutants says: "It is disturbing to think that all of us are being poisoned at least a little bit by lead."

If you are tired or irritable, you may be suffering from a mild case of lead poisoning, he says. If you go to the seashore or to the country, you may feel better. This is because you are getting away from lead pollution. If you play ball, jump rope, skate, or go in for other physical activities, you may also feel better, because perspiration helps remove traces of lead from your body.

Five million children in the United States are considered to be overactive. They have trouble sitting still and reading, studying, building things, or playing quiet games. Some doctors believe that a number of these children may be suffering from lead poisoning. A pediatrician—a children's doctor—compared overactive children with typically active children. The children lived in various New York City neighborhoods. He found that many of the overactive youngsters, not the typically active ones, had large amounts of lead in their blood. He concluded that large amounts of lead caused, or at least contributed to, their overactivity.

Some 200 American children die each year from nibbling bits of lead paint from walls. Some 400,000 other American youngsters get sick from this.

Cattle grazing near highways have collapsed and died from long-range exposure to lead in automobile exhaust. Some scientists considered recycling (reusing) old newspapers by feeding them to cattle. Then they learned that this was not a good idea. The ink on the newspapers contains lead that could poison cattle.

In the Staten Island, New York, zoo, two leopards became paralyzed. A horned owl's feathers dropped out. Snakes lost their ability to

Car exhaust makes up a major part of the smog that has endangered inhabitants of cities for many years. Not only animal life but plant life as well is affected. This petunia plant shows an ailment called ''silvering'' or ''bronzing,'' an effect of smog. The damage appears near the tips of younger leaves and nearer the center and base of the older leaves.

slither. The animals turned out to be suffering from lead poisoning. They had been exposed to traces of lead in grass, leaves, and soil outside their cages and to lead paint on the bars of their cages.

In plants, lead goes for cells in leaves, particularly for chloroplasts. These particles of living substance with chlorophyll in them are the plants' energy factories. Photosynthesis is carried out in them. Traces of lead can tie up enzymes in chloroplasts, disrupting photosynthesis. If plants cannot eat and breathe, they wilt and die.

In animals and people, traces of lead can inhibit enzymes in the mitochondria. These are the energy factories of animals and human cells. Lead can interfere with iron-containing enzymes in blood so that hemoglobin cannot be made. It can attack the nuclei of cells, which contain cells' genetic material. It can inhibit enzymes in liver cells—enzymes needed to use drugs and dispose of them. It can keep helpful trace elements from passing through cell membranes. Lead can destroy a nerve chemical that nerve cells use to communicate with muscle cells.

Depending on how much lead animals and people get in their bodies, they can suffer some or all these consequences: anemia, kidney disease, liver disease, muscle paralysis or overaction, brain damage, convulsions, death.

Lead pollutants, as we know, have hurt human reproduction since Roman times. But scientists are just now finding out the details of the injury. Animal experiments have shown that traces of lead can seriously damage both sperm and eggs. Lead can pass from a mother's blood to the blood of the fetus she carries in her womb, even though the two circulatory systems are separate. If pregnant animals are exposed to high enough levels of lead, their offspring may be born dead or may not grow properly. They also give birth to fewer offspring. Lead can probably do similar damage in human reproduction.

The traits people and animals inherit can make them more vulnerable to lead poisoning. Two pediatricians found that children lacking a particular enzyme in their blood were more open to lead poisoning than were children with the enzyme.

Many of the bacteria people and animals are exposed to through air, food, and water make poisonous substances called endotoxins. Scientists have found that if endotoxins get into the body along with lead, this pollutant is even more harmful than usual. And the lead, under these conditions, appears to attack enzymes.

Lack of enough vitamin C can make one more susceptible to lead poisoning too, it seems. Young guinea pigs were given a diet lacking

LEWIS W. KOSTER

Lead poisoning can affect muscle cells, such as these fibers of striated muscle magnified 800 times, to such an extent that an animal may lose its ability to move.

vitamin C. A second group received food without C but with lead. Others were given a diet with both C and lead. The animals getting C and lead showed no serious signs of disease. Those simply lacking vitamin C had some trouble with their muscles. Those lacking C and also taking in lead completely lost their muscle control and their ability to move. So, as a precaution against lead pollutants, if for no other reason, be sure to get enough vitamin C.

Dangerous Cadmium

Like lead, cadmium is highly toxic and is a widespread trace pollutant. In the opinion of some scientists, this heavy metallic element is the second most dangerous pollutant we are being exposed to.

Cadmium in fertilizers, manure, and roadside soils is easily absorbed through the roots of important food crops. They include wheat, corn, rice, oats, peas, beets, and lettuce. Cadmium moves swiftly from the roots of plants up into their leaves. Plants grown in soils near heavily traveled roads can get cadmium from the air. Rice grown in the cadmium-polluted Jintsu River in Japan had five times more cadmium in it than rice grown elsewhere in Japan.

Some scientists who were interested in using sewage for crop fertilizers worried about the large amounts of cadmium pollutants in the sewage. Sure enough, the crops hungrily took up the cadmium in the sewage.

Cadmium in polluted streams is taken up by snails. Bacteria, fungi, and fish that eat the dead snails absorb it from them. Fish can take in the element directly from cadmium-polluted waters. Cattle grazing on pastures containing cadmium produced milk that also contained it. When it travels from industrial areas through the air to the arctic it settles in lichens on the Arctic tundra. The lichen (two plants tangled together that help each other live) is one of the reindeer's main sources of food. Reindeer get cadmium in their bodies by eating the lichens; then Eskimos are affected by eating the reindeer.

Cadmium pollutants are easier to inhale from the air than they are to take in from food.

People also swallow the substance in drinking water, in which case it comes partly from the water pipes. People who live in houses with older piping may get a lot of cadmium pollutants in this way, especially if they have "soft" water (that is, containing almost no dissolved salts).

Persons who smoke are exposed to 50 times more cadmium pollution than nonsmokers. But even if you sit in the same room with smokers, you are probably being exposed to some of the pollutant.

Traces of cadmium in a person's hair can tell how much cadmium pollution he or she is exposed to. A lot of it was found in samples of hair taken from Americans living in northwestern Indiana. That area of Indiana is heavily industrialized. But there was more of the element in men's hair than in women's; and more in children's hair than in adults'.

If persons are exposed to cadmium pollution throughout their lives, they will accumulate the element in their bodies. One study showed that a 50-year-old woman had 300,000 times more cadmium in her body than her newborn grandchild.

Once cadmium pollutants enter our bodies, they ride piggyback through the bloodstream on proteins called albumins. As more and more cadmium races through the blood, it may settle in blood vessels, liver, kidneys, even in bones. If these tissues and organs get overloaded with it, cadmium can replace traces of zinc in certain enzymes, since cadmium is a heavier relative of zinc. If cell activities in these tissues and organs are hurt enough, one may suffer serious diseases.

Cadmium can cause emphysema, a lung disease, in people who inhale cadmium from cigarette smoke. Cadmium has been linked with cancer; in fact, it appears to be one of the most powerful of all cancer-provoking elements. Just one injection of the heavy metal can cause cancer in experimental animals. Cadmium appears to be a major factor in high blood pressure—which in turn is a major cause of heart disease and stroke (brain hemorrhage).

In 1960 many people who died from heart disease were found to have drunk soft water rather than hard water during their lifetime. Because soft water is rich in cadmium pollutants, scientists believe they may be the cause of high blood pressure. A St. Louis, Missouri study showed that patients with high blood pressure had 50 times more cadmium in their urine than did healthy persons. A study in Czechoslovakia showed that the kidneys of people with high blood pressure were overloaded with cadmium. When rats, rabbits, and dogs were given cadmium, it ended up in their livers, kidneys, and blood vessels. They also got high blood pressure.

Cadmium pollution is widespread in Japan. The Japanese have more cases of high blood pressure than do most people in the world. The leading cause of death in Japan is stroke. On the other hand, some peoples in Africa and Thailand have very low blood pressure, and no cadmium at all can be found in their kidneys.

Is it cadmium in the liver that causes high blood pressure and in turn heart disease or stroke? Or in the kidneys? Or in the blood vessels? Or in all these tissues? No one is sure.

People who live along the Jintsu River in Japan have long been known to have an ailment called "ouch-ouch" disease. They suffer severe pains in their backs, waists, and spines. Scientists now know that ouch-ouch disease is due to their eating cadmium-polluted rice from the river. Apparently the element causes this disease by releasing important traces of calcium from bone. When laboratory animals were given a diet rich in cadmium, their spines became deformed. Cadmium also increased in their bones while calcium decreased.

Just as a lack of vitamin C in your diet can make you more susceptible to lead pollutants, the same holds true for cadmium.

Mercury Pollutes

Mercury is another heavy element that trace-pollution scientists seriously worry about.

Some mercury may get into living organisms

from natural sources. Tons of mercury reach the oceans from the breakdown of rocks and minerals. Museum specimens of tuna caught in 1900 had as much mercury in them as tuna does today. This finding suggests that the incidence of mercury pollution has not risen much during the past century, and that mercury in living organisms is mostly from natural sources.

On the other hand, there is ever-increasing evidence that the mercury getting into plants, animals, and people is from human pollution. Over the past 150 years, birds that prey on fish have been found to have ever-greater amounts of mercury in them. The birds eat fish that live in mercury-polluted lakes, rivers, and streams. Some 83 per cent of 500 samples of water taken near industrial sources were found to be thick with mercury.

Fungicides—chemicals that kill funguses on plants—that contain mercury also appear to be an increasing source of danger. Game birds absorb mercury from fungicide-treated crops. Fish get mercury from fungicides washed into waterways. Birds of prey eat fish contaminated with fungicidal mercury. People eat breads made from grains treated with mercury fungicides. They also eat fish and birds containing mercury fungicides.

Even if mercury pollutants are not toxic,

The hatching of a quail chick never does occur in places where pollution by mercury compounds causes the shells to break before the quail has developed sufficiently.

bacteria can convert them to toxic forms. Mercury has damaged the leaves of lettuce and carrots, stunted the growth of plants, and caused quail eggs to break, disrupting quail reproduction. The major victims of mercury pollution, though, appear to be people. The reason may be that mercury tends to accumulate in organisms as it moves up the food chain.

The first known examples of widespread mercury poisoning occurred in the 1950s. Forty-six persons died, and 65 were seriously sickened by eating shellfish taken from Japan's Minamata Bay. The shellfish were filled with mercury pollutants discharged by a plastics factory on the bay. During the 1960s and early 1970s, thousands of people in Iraq, Pakistan, and Guatemala were killed or seriously harmed by eating flour and wheat seeds treated with fungicidal mercury.

Apparently it took only a few traces of mercury in the environment to produce these results.

As with lead poisoning, mercury poisoning takes place over many weeks and months. During this time, symptoms are hardly apparent. Then as traces of the element accumulate in the body, they combine especially with blood, nerve, and liver cells. Once traces of mercury are inside these cells, they disrupt certain kinds of proteins and damage chromosomes.

When enough traces of mercury build up in a person's body, he or she is poisoned. Mercury poisoning can lead to a loss of the sense of touch in hands and feet. It can cause a loss of hearing and vision. If a person is poisoned enough, he or she can pass into a coma and die.

Mercury is able to pass from the womb of a woman to the fetus she carries and harm it. Fetuses poisoned by mercury in the womb have been born mentally retarded. Some have also been born with cerebral palsy—a disease that is the result of brain damage and disrupts the normal motions of muscles.

Pollution From All Sides

Traces of the middle-weight element nickel can be harmful. In Scotland and southern Rhodesia, soils that are naturally high in nickel have

THE INTERNATIONAL NICKEL COMPANY

A "mucker" lifting thousands of pounds of nickel ore. An excess of nickel in soil can prevent plants from reproducing, and a nickel compound is one cause of lung cancer.

kept plants from reproducing. Since nickel is present in pesticides, fuels, and tobacco smoke, these are among the sources. A compound known as nickel carbonyl is formed when the element is emitted during the burning of fuels. Nickel carbonyl has caused lung cancer in some industrial workers, also in laboratory animals. Traces of nickel in asbestos may cause lung cancer in asbestos workers.

Asbestos is a chemical compound. The commonest form is made of hydrogen, magnesium, silicon, and oxygen. It is made into fibers that are used to fireproof and insulate houses, apartments, and offices. Asbestos is used to line car brakes and as paint filler. Scientists have found that asbestos dust is particularly rich in traces of nickel. They are now trying to see whether nickel from asbestos dust causes cancer in laboratory animals.

Nickel screws and rods are sometimes implanted in the legs and hips of people who need leg or hip support. Some persons have been infected by such nickel implants, so the implants had to be removed.

Of all the elements, the fourth lightest element—beryllium—is the most toxic. Scientists don't know why. Workers who get traces of beryllium in their lungs develop lung inflammation, sometimes cancer. But beryllium probably does not hurt most people. This is because it is not a widespread pollutant. The only people likely to inhale beryllium are those who work in the beryllium and coal industries, or at Cape Kennedy, Florida. Beryllium escapes from rocket fuels at Cape Kennedy when spacecraft are shot to the moon and other planets.

Traces of the heavy element antimony escape from fumes in printing plants. People who set type in these plants have a high incidence of heart attacks. When traces of antimony were fed to laboratory animals over a lifetime, the traces produced heart attacks in a third of them. So here was more evidence that traces of antimony might cause heart attacks. The big question, of course, is whether this is a widespread pollutant. One pollution scientist doubts that it is.

Traces of middle-weight arsenic are highly poisonous. In Taiwan, people who drank water from a particular well got skin cancer. It turned out that the water contained small amounts of arsenic. One child living on a ranch in Nevada developed skin ailments. Then another child did as well. Their skin conditions were traced to arsenic in water they drank on the ranch.

Bits of tungsten and barium, both heavy elements, have reduced the life expectancy of experimental animals. Yet trace pollutants of tungsten, largely from steel industries, probably do not hurt the general public. Nor do traces of barium pose a general threat. Barium is naturally present in the environment in large amounts. So plants, animals, and people are probably used to it. In addition, people who work around barium have not experienced any serious health problems.

Although phosphorus is one of the six light elements that make up the bulk of living organisms, trace pollutants of phosphorus can have one bad effect at least. Phosphorus is in detergents. If detergents are dumped into waterways, they cause an overgrowth of undesirable algae, or water plants. Such algae have spread rapidly throughout the Great Lakes.

Traces of the middle-weight element zinc are helpful, but a little is enough; beyond a certain point it can be harmful. Zinc pollutants from zinc smelters, industrial wastes, pesticides, and sewage can hurt plants.

Palmerton is a small town in Pennsylvania. Two zinc smelters operate there, giving off a lot of zinc pollutants. A biologist found that trees near the smelter had a lot of zinc in them; they were scrubby and their leaves were red, as if autumn had already come. These effects were apparently due to the zinc. Trees farther from the smelters had much less zinc in them and they looked a lot healthier.

Copper is another helpful middle-weight trace element. Yet it too may became a hazard under certain conditions. Traces of copper in fungicides have hurt orange groves in Florida, vineyards in France and Italy, and apple orchards in England. Sheep are more susceptible to copper poisoning than are other domestic animals. In the Netherlands, traces of copper escaped from power lines and polluted the soil underneath them. Plants soaked up the excess copper from the soil; sheep ate the plants and died.

The heavy element silver is not a widespread pollutant. Yet traces of it escape from silver mines in the West, enter nearby streams and kill trout.

As we can see, scientists have a fairly good idea which trace pollutants are harmful and which are not. But they still have some important questions to answer. They must monitor various areas for all kinds of trace pollutants. They must screen different kinds of foods and check drinking water. They must learn how much of various elements can be taken into a plant, animal, or person before they do damage. Scientists are finding that there is often a slim margin between traces of an element that are helpful and traces that are harmful. Elements are something like banana splits. A little can be delicious; a little too much can give one a stomach ache. Only when scientists get answers to these questions will they be able to say with certainty which trace pollutants are hurting whom and how to protect us.

What's Being Done?

Meanwhile, some protection efforts are being made. Since 1971, the Environmental Protection Administration of New York City has required a steady reduction of lead in gasoline year by year. The United States Environmental Protection Agency has ruled that after July 1974 at least one grade of lead-free gasoline must be sold throughout the United States. Many automobiles, to meet official 1975 standards on what car exhaust may contain, will have to use lead-free gasoline.

A few cadmium industries have reduced

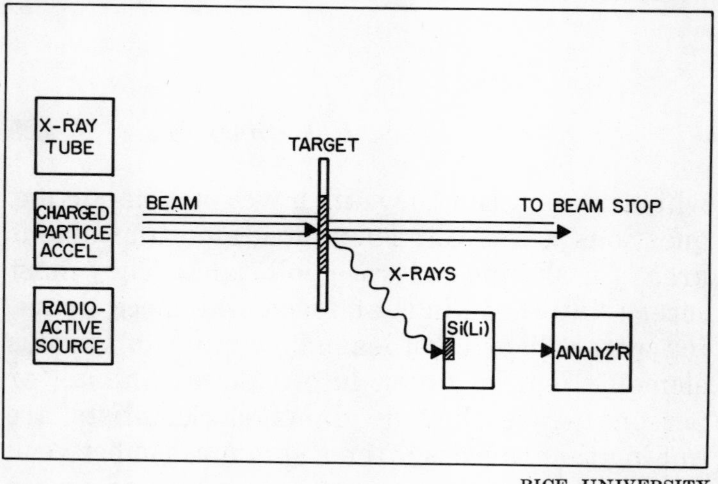

RICE UNIVERSITY

An X-ray spectrometer can identify a trace element. This generalized diagram shows how a beam of either X rays or charged particles hits the target, which is the substance containing the unknown element; the target then gives off at an angle X rays that are characteristic of the substance. The Si(Li) detector measures the energy of each X ray reaching it, and this identifies the element by showing its atomic number.

their cadmium pollutants. Unfortunately the Environmental Protection Agency has not yet set any requirements for the reduction of cadmium pollution. Industries are working on ways to avoid mercury pollution, and are dredging mercury pollutants from waters. They are trying to pin down other trace pollutants they create, so they can do something about them.

Scientists at Rome, New York, made a major effort to identify the sources of mercury pollutants and their effects on wildlife. As a result of large-scale testing of fish, they estimate that industrial mercury discharges into New York waterways have been reduced by 97 per cent. This is real progress.

Canada and the United States, after six years of study, have decided to limit dumping of phosphorus into the Great Lakes. Phosphorus is to be limited to 16,000 tons in 1976. Canada has already ordered that phosphorus in detergents be reduced.

A soil chemist wanted to keep trace elements in fertilizers from polluting the environment. She devised a system that indicates precisely various crops' needs for certain trace elements. These include nitrogen, phosphorus, potassium, calcium, magnesium, manganese, copper, zinc. Once the amounts needed are known, perhaps fertilizers can be made without excesses of them.

Michigan chemists looked for ways to keep cadmium pollutants from flaking off water pipes into drinking water. They found that corrosion of the pipes and entry of cadmium into drinking water could be reduced by coating the inside of the pipes. One of the chemicals they use for this is carbonate. Hard water contains a lot of carbonate. This probably explains why certain hard waters have helped keep cadmium out of drinking water.

Dialysis is a technique with which impurities are removed from the blood of patients with kidney disease. Some New York scientists used dialysis to clear mercury from the blood of experimental animals. They found that they could remove this element from the animals' blood much faster than the blood would normally clear it. The investigators think the technique might be used on persons who are poisoned by mercury or other heavy elements. So, step by step, in laboratories, offices of antipollution organizations, and legislative halls, protection is beginning to come —but as yet far too slowly.

10/ Helping Trace Elements Help People

The scene is a trace-elements laboratory in Maryland. Scientists in white coats are pouring bubbling liquids into large flasks. They are turning dials and pulling levers on impressive and expensive machines. With one push of a button on a machine called a chromatograph, they can split the element chromium into a million traces. Says one of the scientists, "It's like setting off a rocket to the moon."

These scientists, like some others throughout the United States and in other countries, are busy studying the role of trace elements in medical treatment. This is an area that researchers have not paid much attention to up to now. This particular group is trying to get a better idea of the role of chromium traces in sugar metabolism —the way the body breaks down and uses sugar.

They have found that traces of chromium

are important ingredients in a chemical material called the "glucose-tolerance factor." The factor is a substance that our bodies need to use sugar properly. Many people do not break down sugar well as they grow older. Also, traces of chromium in their tissues decrease. So possibly, the scientists theorize, these people need dietary supplements of chromium. Then their bodies might be able to use sugar better.

Gold and Arthritis

Traces of gold have long been used to treat arthritis. The gold slows down enzymes that chew away at cartilage and bone. But sometimes patients get too much gold in their bodies from this therapy and show signs of heavy-element poisoning. Now California scientists have found a way of making gold treatment more effective and safe.

Instead of injecting patients with a standard dose of gold, they carefully measured the amount of gold that appeared in patients' blood. They found that different patients used gold at different rates. They tried adjusting people's weekly doses of gold so that its level in their blood remained constant. This means that sometimes a patient got more gold than the standard dose, sometimes less, sometimes none at all. The patients did well with this new, measured treat-

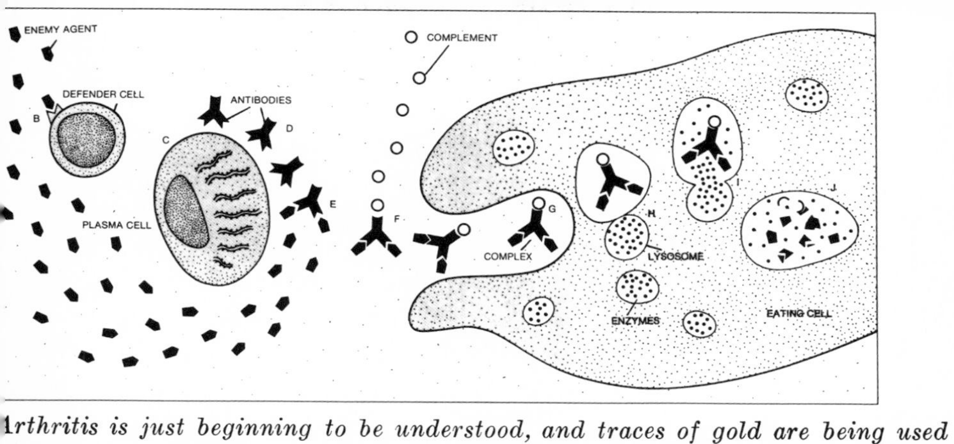

*rthritis is just beginning to be understood, and traces of gold are being used ith partial success in treating it. For comparison with the following picture, his is what happens when "enemy agents," such as destructive bacteria or iruses, attack the body and the body responds normally to put them out of ommission. An enemy agent (*A*) is intercepted by a defender blood cell (*B*) hich is stimulated (*C*) into producing protective antibodies (*D*). The anti-odies attach chemically to the enemy agents (*E*), and attract and join with omplements—additional chemicals (*F*). This chemical group is swallowed by a ody cell (*G*) and brought into contact with lysosomes, small sacs full of en-ymes that dissolve the enemy agents captured by the antibodies.* (BOTH PICTURES, HE ARTHRITIS FOUNDATION, *1969 Annual Report*)

---

n arthritis—which researchers suspect may be caused by a yet unknown virus—he normal defense system of the body described in the preceding picture is hrown off its normal course and itself attacks the body. The result is inflamma-ion of the joints, which is arthritis. At left, a "confused" defender blood cell ttacks a cell of the synovial lining that produces synovia, a lubricant that eases he motion of joints. At right, as the complexes—the chemical groups each onsisting of enemy agent, antibody, and complement—are taken in by the ating cell, it becomes gorged and lets out powerful enzymes that escape into he joint lubricant and damage joint tissue. It is these enzymes that traces of old can hamper to a certain extent.

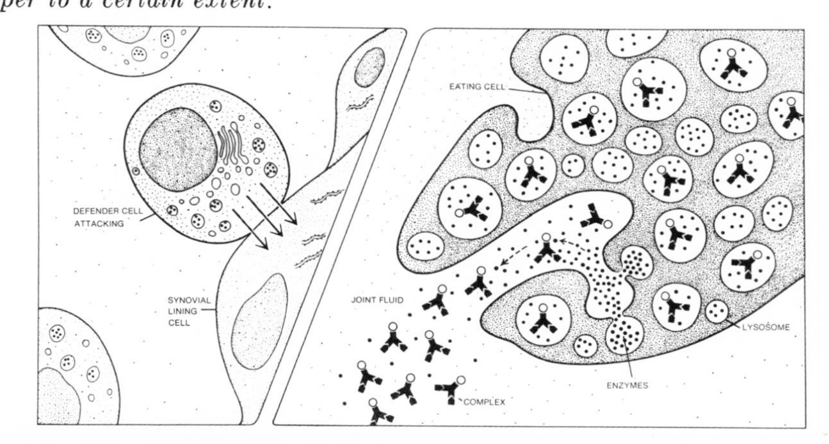

ment. Not one of them suffered from gold poisoning.

The test also produced unexpected results. One patient thought that if a little gold is good, more of it is better. He had been getting shots of gold from more than one physician. He was found out because the level of gold in his blood was high, even when the scientists withheld gold from him.

Still other scientists have reported that loss of appetite, and a changed sense of taste, in patients who have been severely burned is due to a loss of zinc from their bodies. They conducted tests on 19 patients, all suffering from burns. Each patient was tested for the sense of taste in four qualities—salty, sweet, sour, and bitter. Sixteen of the 19 suffered from loss of taste. The patients were then tested for the content of zinc in their bodies, and their levels were below normal. The investigators are now going to give supplements of zinc to burned patients to see if the traces correct their loss of appetite and changed taste.

The silvery gray, middle-weight element titanium looks as if it will be of value in leg supports. As mentioned earlier, some patients have become infected from nickel screws and rods implanted in their legs. The reason is that nickel is broken down by chemicals in the body. Titanium appears to resist such destruction.

The value of trace elements in medical diagnosis has been largely ignored. But some scientists are now paying more attention to this vital field. For example, some researchers have noted that there are unusually large amounts of copper in the fingernails of infants with cystic fibrosis, an inherited disease that causes difficulty in breathing and in clearing mucus from the lungs. So the researchers have developed a means of diagnosing trace levels of copper in newborns, and hope that their technique will be used to mass-screen newborns for the disease.

California biologists have found that heavy trace elements such as nickel and cadmium, known to cause cancer in experimental animals, are carried in the bloodstream by albumins. But noncancerous trace elements such as iron and zinc are carried by globulins. Albumins and globulins are proteins in the blood. The researchers hope that their findings may be a way of telling which trace elements cause cancer in people and which do not. To find out for sure, they are now testing other trace elements, to see which are carried by albumins and which by globulins.

Diagnosing by Trace Elements

Few physicians have their patients analyzed for the trace elements. Yet there is more and more evidence that such analysis is useful in medical diagnosis. An Ohio scientist has found that stresses, such as operations and injuries, increase traces of zinc in people's blood. The amount of zinc that shows up depends on how severe the stress is. The increase goes away when the stress lets up. She suggests that analysis of zinc in blood might provide a means of monitoring the recovery of critically ill patients.

A Russian scientist reports that the amounts of trace elements in blood may be used to diagnose certain types of heart disease. He has found that traces of nickel and manganese rise in the blood just before heart attacks. He therefore suggests that patients with heart disease have their blood analyzed periodically for any rises in nickel and manganese.

Other diseases as well can be predicted. In the past, scientists have been able to detect only one trace element at a time in samples of blood or tissue. After nine years of work, a chemist has developed a technique that can determine 60 trace elements in a sample of material within one minute.

A sample of blood, say, is placed in the bottom of a device that looks like a two-inch torch. The blood is sprayed up the torch into a "plasma." The plasma looks like flame shooting from the end of the torch. But it is a gas that is hot enough to shatter trace elements into even smaller pieces. As the trace elements shatter,

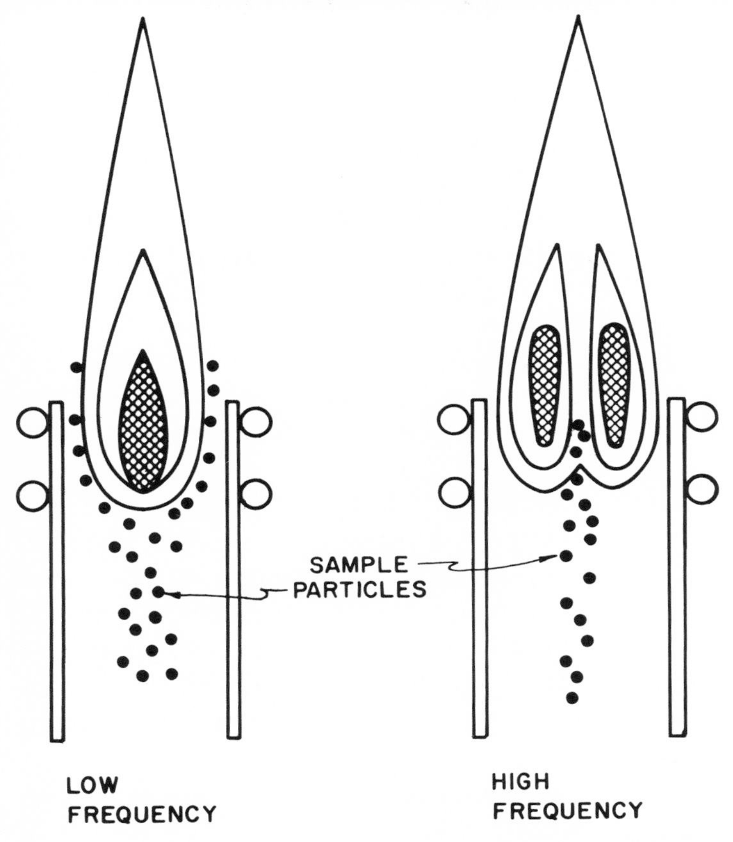

AMES LABORATORY, AEC

Two forms of plasmas, or "electrical flames." A plasma is any luminous gas in which a certain portion of its atoms or molecules are ionized—that is, have lost or gained one or more electrons, and therefore carry an electrical charge. An element that is vaporized by being placed in a flame, such as the gas flame of a Bunsen burner, gives off its own kind of light, and an optical instrument called a spectroscope can reveal what element it is from this light. Such a method of analysis has long been in use. A plasma "electrical flame" can do the same, but used in connection with a computer it gives greater sensitivity and accuracy in detecting trace elements. Drawing originally published in Analytical Chemistry.

they give off rays of light. Each kind of element gives off a particular wavelength of ray. A computer attached to an instrument called a spectrometer measures both the wavelengths and the intensities of the rays. The wavelength shows which trace element is in the sample; the intensity shows how much of each trace element is in it.

Other scientists are finding ways of using trace elements in disease prevention. For instance, trace elements can help protect our teeth from decay. Traces of fluorine were first put in drinking water in the United States in 1945. Research indicated that traces of fluorine drastically reduce tooth decay in young people. During the years following 1945, fluoridation of water spread to thousands of American communities. More recently, investigators have found that not just fluorine but also traces of some other elements can protect teeth from decay. They are aluminum, iron, and tin.

Animal experiments have shown that fluorine's protection is influenced by traces of iron or aluminum in drinking water. Chemical analyses actually showed that aluminum and iron, as well as fluorine, enter the enamel of teeth. Fortunately aluminum and iron are often added to drinking water by city treatment plants, and traces of tin are present in toothpaste.

Solving Crimes

While some scientists are looking into the value of trace elements to medical diagnosis, treatment, and prevention, others are examining their value to forensic medicine—the branch of medicine that helps detectives solve crimes.

This specialty is opening the exciting possibility of tracking down criminals through trace elements present in hair that is left near the scene of a crime. How is this possible? If trace elements in the hair samples are similar to those in the hair of a suspect, the possibility that the suspect committed the crime is strengthened. Or if a hair sample has certain traces in it, a detective might figure out who the criminal is because he knows that workers from a certain industry or residential area have the same trace elements in their hair.

Before a detective knows which types of people carry along certain substances on them, though, scientists must get trace-element profiles on many people. Certainly one's work and location decide the traces one carries. But other factors can decide their presence as well.

Chemists recently conducted a study of trace elements in hair from many people. Some of their samples were even historical, coming from no less famous persons than George and Martha Washington and Abraham Lincoln. They found that two major factors can decide what trace elements end up in hair: sex and diet.

Girls have more aluminum, calcium, and iodine in their hair than boys. Boys have more manganese and copper in their hair than girls. Eskimos have more magnesium, manganese, arsenic, iodine, and mercury in their hair than do midshipmen at the United States Naval Academy in Annapolis, Maryland. But the midshipmen have more calcium, chromium, and possibly iron. Apparently the Eskimos' trace elements come from a diet that is high in salt-water fish. The midshipmen probably get their calcium from drinking a lot of milk. The chromium may come from eating food that had been wrapped in packages containing chromium traces.

The chemists found some other surprising factors that can influence what is found in hair. If you use shampoo with traces of zinc in it, you will have slightly more zinc in your hair than otherwise. If you swim in a pool of water that contains the liquid red element bromine, a lot of bromine will end up in your hair. Traces of gold or silver in your hair probably come from gold or silver caps on your teeth.

Trace elements are also helping the science of archaeology. Geologists are working on identifying objects from the Bronze Age through analysis of trace elements. A major one in ornaments from this period is copper. So the scientists reasoned that the copper in necklaces, urns, and other treasures might show what ancient peoples made them, by indicating where the copper came from. In fact, some of the specimens were traced to the very mines they were taken from.

A Michigan chemist can tell which silver coins are ancient and which are modern imitations. The ancient silver contains traces of gold, but not modern silver. He found that even some silver coins in the most famous museums in the world are fakes. He also found that a French coin saying "1578" was not an authentic silver coin from that period. The real coins were 30 per cent silver. But the fake coin was 75 per cent copper and 25 per cent arsenic.

He and his wife also shed light on an economic inflation that swept Europe in the sixteenth and seventeenth centuries. The inflation was supposed to have resulted from the Spaniards' bringing Mexican coins to Europe. But chemical analysis showed that the Mexican coins had different amounts of gold in them than did coins used in Europe during the inflation. So the chemist and his wife conclude: "The historic basis for this theory of inflation is incorrect. This chapter in history books has to be revised."

Trace elements are also helping art. They are

being used to expose paintings that are imitations of originals. If a painting contains traces of cobalt, say, and cobalt was not in paint pigments used at the time the original painting was made, then the painting with cobalt is obviously a fake.

Lab Animals Answer Medical Questions

Probably the greatest progress scientists are making in helping trace elements help people is learning more about which ones are essential to our health. Because they can't very well experiment on people, they use laboratory animals. Whatever results they get with animals can sometimes be applied fairly safely to people, though there are many exceptions.

First the animals are put in cages that are absolutely free of trace contaminants. The cages are made entirely out of plastic, even the food cups and water bottles. The cages have air locks on them. Once test animals are put in the cages, they are fed highly purified diets containing all the nutrients the animals need except for the trace element being tested. If the animals suffer some defect in health that does not seem to have any other cause, then the element is probably essential to their health. If they are then given the trace element and the health effects are reversed, the element is definitely shown to be essential.

These elegant research techniques have

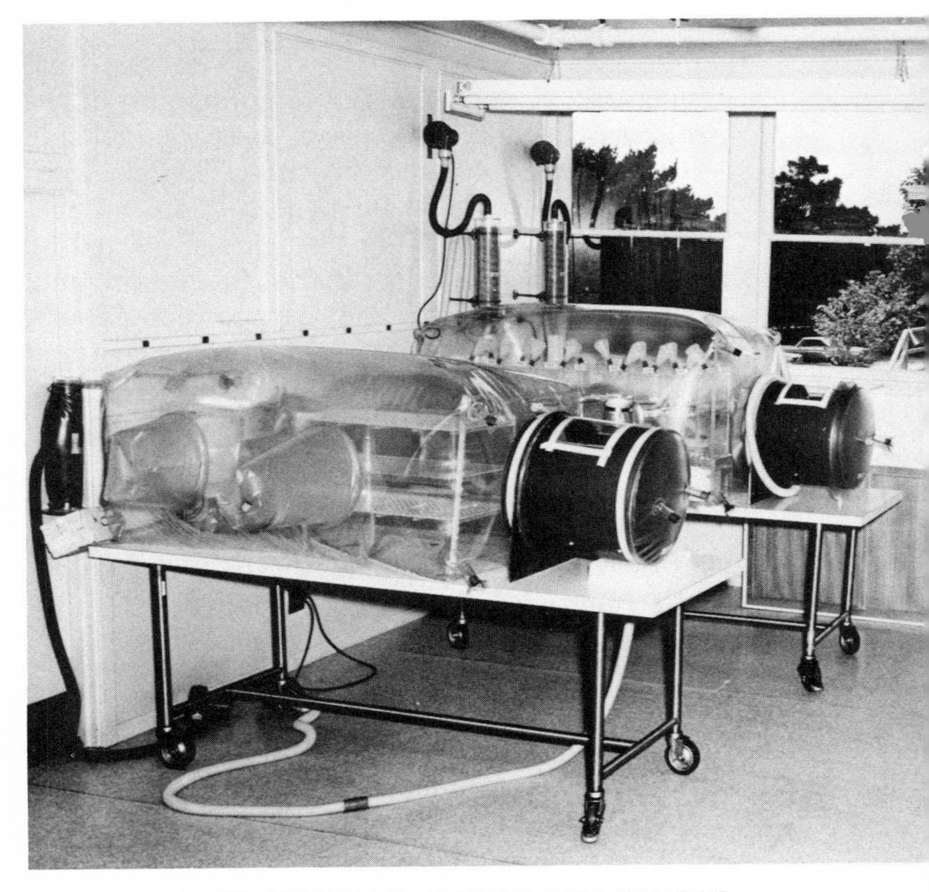

DR. KLAUS SCHWARZ, VETERANS ADMINISTRATION

A closer view of the plastic sterile isolators seen in the background of the frontispiece. Environments provided by such isolators are important not only to trace-element research but also to investigation of many diseases and of the body's defense systems. Raising animals free of all bacteria and viruses in such cages enables scientists to make tests never possible before they were invented.

been available only about 20 years, so most of the progress in identifying essential trace elements has taken place during that time. The ones considered essenital now include: sodium, magnesium, potassium, calcium (which are needed by animals and people in somewhat larger amounts than traces); iron, iodine, copper, manganese, zinc, cobalt, molybdenum, selenium, chromium, tin, vanadium, fluorine, nickel. The one to emerge most recently as essential is silicon, a nonmetallic light element, very common in the earth's crust, and always combined with oxygen.

California biological chemists found in 1973 that traces of silicon help maintain the structure and strength of skin, cartilage, ligaments, and other connective tissue—that which supports and binds together the parts of the body. When rats were fed diets lacking silicon, they did not grow properly and developed badly shaped bones. The same thing happened to chickens.

As things look now, six light elements account for the greater part of animal and human tissue. Other elements are essential in small or trace amounts. All of these are light or middleweight, except for iodine and tin, which are heavy. The question is: will more light and middle-weight elements be found essential in trace amounts? There are some possibilities—arsenic, for example, even though it is deadly if much is given to a person. Even more intriguing: might

DR. KLAUS SCHWARZ, VETERANS ADMINISTRATION

The animal at top was kept in an ordinary cage under usual conditions in which various trace elements were available. The one below was fed for 20 days on the same diet but was kept in an isolator free of all trace elements in its surroundings. It suffers from trace-element deficiency.

more heavier elements turn out to be essential in trace amounts? Traces of some of them, such as mercury, cadmium, lead, antimony, silver, and gold, are found in people in more or less constant amounts. The quantities can, of course, be altered by trace pollutants.

Some scientists believe that some heavier elements will turn out to be essential in extremely small amounts. Although trace quantities of these elements can be dangerous to health, even smaller amounts might have some beneficial effect. Minute quantities of some heavy elements, such as lead, have been found to promote growth in animals. Other scientists, however, are convinced that the heavy elements will not turn out to be essential.

What scientists do tend to agree on, though, is that more trace elements await discovery as dietary essentials. Will they be indium, palladium, gallium, rubidium, or other middle-weight elements that we don't hear much about?

Meanwhile, another big challenge faces the investigators. It is figuring out how much of the essential trace elements people need in their diets every day. Then scientists can do their best to make sure that we get these trace elements in foods, or—for those who can afford them—in vitamin-mineral pills. The minimum daily requirements for only a small number of trace elements are now known.

Such research moves slowly. But if you are lucky, one day you may be able to make sure that you get all the essential trace elements that you need every day. You may decide on copper for breakfast, tin for lunch and—who knows?—chromium for dinner!

Glossary

aluminum A bluish-silvery white, metallic element. Light. Atomic number 13 in the Element Table. Important to some plants. Its value to animals and people not yet proven.

antimony A silvery-white, metallic element. Heavy. Atomic number 51 in the Element Table.

argon A colorless, odorless, gaseous element. Light. Atomic number 18 in the Element Table.

arsenic A steel-gray, middle-weight element. Number 33 in the Element Table.

atom The smallest quantity or unit of any element that enters into chemical combination.

atomic number The number of protons in the nucleus of an atom (as well as the number of electrons outside). Lithium has 3 of each, so its atomic number is 3.

barium A steel-gray, metallic element. Middle weight. Number 56 in the Element Table.

beryllium A steel-gray metallic element. Very light. Number 4 in the Element Table.

bismuth A grayish white, metallic element. Very heavy. Number 83 in the Element Table.

boron Neither metallic nor nonmetallic. Very light. Number 5 in the Element Table. Important to plants, apparently not to animals and people.

bromine A deep red, liquid, middle-weight element. Number 35 in the Element Table.

cadmium A silvery-white, metallic, heavy element. Number 48 in the Element Table.

calcium A silvery-white, metallic element. Middle weight. Number 20 in the Element Table. Important to plants, animals, and people in trace, or very small, amounts.

carbon A nonmetallic, very light element. Number 6 in the Element Table. Found in coal, graphite, diamonds, limestone and other substances that make up the earth's crust. One of the six elements that make up the greater part of tissues.

cesium A silvery-white, metallic element. Rather heavy. Number 55 in the Element Table.

chlorine A greenish-yellow, gaseous element, of pungent odor. Light. Number 17 in the Element Table.

chromium A bluish-white, metallic element. Middle weight. Number 24 in the Element Table. Important to animals and people in trace amounts.

cobalt A lustrous, silvery white element. Middle weight. Number 27 in the Element Table. Important to animals and people in trace amounts.

compound A pure substance made of chemically combined elements.

copper A reddish, metallic, middle-weight element. Number 29 in the Element Table. Important to plants, animals, and people in trace amounts.

electron A unit of negative electricity in an orbit around the nucleus of an atom.

element The simplest form of matter; each piece of an element contains only atoms of the same kind, with the same atomic number.

fluorine A pale yellow, gaseous, light element. Number 9 in the Element Table. Important to animals and people in trace amounts.

gallium A bluish-white, metallic, middle-weight element. Number 31 in the Element Table.

germanium A grayish-white, middle-weight element, neither metal nor nonmetal. Number 32 in the Element Table.

gold A lustrous, yellow, heavy element. Number 79 in the Element Table.

helium A light, colorless, gaseous element. Number 2 in the Element Table. With hydrogen, it makes up most of the universe.

hydrogen A colorless, odorless, gaseous element. Simplest and lightest of all the elements. Number 1 in the Element Table. With helium, it makes up most of the universe. Occurs in combination with other elements more than any other element. One of the six elements that make up the greater part of tissues.

indium A silvery, metallic element. Middle weight. Number 49 in the Element Table.

iodine A blackish-gray, nonmetallic element. Heavy. Number 53 in the Element Table.

Important to animals and people in trace amounts.

iron A silvery white, metallic element. Middle weight. Number 26 in the Element Table. Important to plants, animals, and people in trace amounts.

lead A bluish-white metallic element. Very heavy. Number 82 in the Element Table.

magnesium A silvery white, metallic, light element. Number 12 in the Element Table. Important to plants, animals, and people in trace amounts.

manganese A grayish-white, middle-weight element. Number 25 in the Element Table. Important to plants, animals, and people in trace amounts.

mercury A silvery white, metallic element. Very heavy. Number 80 in the Element Table. Ordinarily liquid.

mixture A combination of substances held together physically rather than chemically.

molecule The smallest particle of a compound that shows the characteristics of the compound. Made of combined atoms, except for a few cases in which an atom is also a molecule.

molybdenum A bluish-white, metallic element. Middle weight. Number 42 in the

Element Table. Important to plants, animals, and people in trace amounts.

nickel A silvery white, middle-weight element. Number 28 in the Element Table. Important to animals and people in trace amounts.

nitrogen A colorless, tasteless, odorless, gaseous element. Very light. Number 7 in the Element Table. Makes up 78 per cent of the earth's atmosphere. One of the six elements that make up the greater part of tissues.

oxygen A colorless, tasteless, odorless, gaseous element. Very light. Number 8 in the Element Table. Makes up one-fifth of the earth's atmosphere; is the most abundant element in the earth's crust. It is one of the six elements that make up the greater part of tissues.

palladium A silvery white, metallic element. Middle weight. Number 46 in the Element Table.

phosphorus A nonmetallic, light element. Number 15 in the Element Table. Widely found in rocks and minerals. One of the six elements that make up the greater part of tissues.

platinum A grayish-white, metallic element. Heavy. Number 78 in the Element Table.

potassium A silvery white, metallic element. Middle weight. Number 19 in the Element Table. Important to plants, animals, and people in trace, or small, amounts.

proton A unit of positive electricity in the center, or nucleus, of an atom.

rubidium A silvery, metallic, middle-weight element. Number 37 in the Element Table.

selenium A nonmetallic, middle-weight element. Number 34 in the Element Table. Important to animals and people in trace amounts. May help plants.

silicon A nonmetallic, light element. Number 14 in the Element Table. After oxygen, the second-most abundant element in the earth's crust. Emerging as an important trace element for animals and people.

silver A bluish-white, metallic, heavy element. Number 47 in the Element Table.

sodium A silvery white, waxy, light element. Number 11 in the Element Table. Important to animals and people in small amounts.

strontium A metallic, middle-weight element. Number 38 in the Element Table. Important to the oyster in trace amounts, perhaps to other living things as well.

sulfur A nonmetallic, light element, of unpleasant odor. Number 16 in the Element Table. Found in sediment, rocks, and volcanoes. One of the six elements that make up the greater part of tissues.

thorium One of the heaviest elements and rare on earth. Number 90 in the Element Table.

tin A lustrous, bluish-white element. Heavy. Number 50 in the Element Table. Important to animals and people in trace amounts.

titanium A silvery gray, metallic element. Middle weight. Number 22 in the Element Table.

vanadium A gray, metallic element. Middle weight. Number 23 in the Element Table. Important to people, animals, and at least some plants in trace amounts.

zinc A bluish-white, middle-weight element. Number 30 in the Element Table. Important to plants, animals, and people in trace amounts.

zirconium A steel-gray, metallic, middle-weight element. Number 40 in the Element Table.

Suggested Reading

Younger Books

William Bixby, *A World You Can Live In* (David McKay, 1971). A general discussion of the pollution problem, but specific mention of mercury and mine pollution.

Sarah M. Elliott, *Our Dirty Air* (Julian Messner, 1971). Mostly about pollutants that are compounds of elements rather than those that are traces of individual elements. Some comments about lead and fluorine.

Winfred B. Luhrmann, *The First Book of Gold* (Franklin Watts, 1968). Describes how gold is mined and refined, its uses in ancient civilizations, the great gold rushes, gold in the world today.

Patricia Maloney Markun, *The First Book of Mining* (Franklin Watts, 1959). Historical and modern perspective on mining of silver, sulfur, magnesium, gold, lead, iron, tin, copper, cadmium, and other elements.

John Perry, *Our Polluted World: Can Man Survive?* (Franklin Watts, 1967). A general look at the pollution problem, how pollutants hurt plants, animals, and people. Some discussion of lead pollution and lead poisoning.

Marjorie Quennell and C. H. B. Quennell, *Everyday Life in Prehistoric Times* (G. P. Putnam, 1959). Life in the New Stone, Bronze, and Early Iron Ages, including use of the elements.

Beulah Tannenbaum and Myra Stillman, *Understanding Food: The Chemistry of Nutrition* (McGraw-Hill, 1962). Discusses food fads, fancies, fallacies, body chemistry, foods of the future. A chapter on trace elements—which foods they are present in and what they do for your body.

Edward B. Tracy, *The New World of Copper* (Dodd, Mead, 1964). The history of copper, mining, uses, copper in agriculture and health, copper in the future.

Older and Adult Books

Pauline Arnold and Percival White, *Food Facts for Young People* (Holiday House, 1968). Minerals, vitamins, trace elements, and other aspects of food.

J. Calvin Giddings, *Our Chemical Environment* (Canfield Press, 1972). Section IV is devoted to contamination by lead, cadmium, mercury.

Henry Schroeder, *The Trace Elements and Man* (Devin, 1973). Trace-element roundup by one of America's leading trace-element scientists.

Karl H. Schutte, *The Biology of the Trace Elements and Their Role in Nutrition* (Lippincott, 1964).

Howard E. Smith, Jr., *From Under the Earth: America's Metals, Fuels, and Minerals* (Harcourt, 1967).

Walter Stiles, *Trace Elements in Plants* (University Press, Cambridge, Engl., 1961).

E. J. Underwood, *Trace Elements in Human and Animal Nutrition* (Academic Press, 1962).

Magazine Articles

"On the Trail of Heavy Metals in Ecosystems." *Science News,* Sept. 11, 1971.

"Sounding Out Metal Toxicity." *Science News,* Sept. 30, 1972.

"The Hazards of Trace Elements." *Science News,* June 6, 1970.

"Trace Elements: A Growing Appreciation of Their Effects on Man." *Science,* July 20, 1973.

"Trace Elements: No Longer Good versus Bad." *Science News,* Aug. 14, 1971.

Index

DATE